Teaching mathematics to students with visual impairment.

PA

Mathematics teacher I invite you to put yourself in the same shoes as a visually impaired student and take trainings and courses, in order to create more didactic materials. Take courses, have more math books in Braille at all levels since mathematics is adapted but not researched.

Dedicated to the following institutions:

Ezequiel Hernández Romo Institute for the Blind and Visually Impaired.

Faculty of Sciences SLP.

Benemérita y Centenario, normal school of SLP.

Autónoma de México in the Faculty of Sciences.

ONCE.

Instituto Tecnológico de Estudios Superiores de Monterrey.

The Secretary of Public Education (SEP)

   Dedicated to the following people.

Alejandro Fernández Montiel.

To all the mathematics teachers of the science faculty in SLP.

To scientists.

To science researchers.

For students with visual impairment.

Family, friends, acquaintances, among others.

Written by: Omar García Trujillo

Objective of the book:

The purpose of this book is for the mathematics teacher in regular schools to put himself in the shoes of the visually impaired student and to understand that he has difficulties in learning mathematics content at the same pace as a normal-visual student, since the teacher must provide the possible tools to the visually impaired student, but the student must also do his part, so that the teacher does not give him anything for free but supports him; The teacher should be prepared because you never know if you will have a student with visual impairment in your class so it is very important that the teacher has adequate preparation in case someday a student with these characteristics is presented, so it is important training, courses and other things to teachers.

Book introduction:

This book is about mathematics where topics of this subject that is of utmost importance for everyday life will be addressed, so it will also be in Braille as in common physical character in the bookstore as well as in digital and will be in the following languages which are Spanish, English, French and also in Italian, so I recommend you to have it in your home and support us with its dissemination.

So it will have 30 chapters that are very comprehensive as they offer the development of exercises and also has many examples of the topics before you go to check the exercises as the same examples found in each topic were proposed by the author of this book as well as the resolution of these exercises and problems that you will find in this book so peculiar, the topics covered in this book are basic topics as there are also topics that are very complicated to people so this book was designed to help people with visual impairment to learn better math topics.

If you are a math teacher or a person of the common society, encourage students with visual impairment to learn mathematics better, because the visually impaired student has the ability to learn mathematics, but they usually have several difficulties in their learning process, so you are invited to take a little awareness and if you really want to support a student with visual impairment to learn mathematics in the best way and be successful in this field; What are you waiting for? Start training if you are a math teacher, so you are also invited to become aware and put yourself in the shoes of a visually impaired student at least one day of your life. If you are a math teacher who has worked with a visually impaired student at some point, you will understand what I am talking about; If you are a math teacher at whatever level you are, don't be a selfish teacher who only thinks about his miserable salary and if the visually impaired student doesn't care if he learns or not, find a way to help him and if you don't know how to support him, look for support in other countries where there are math teachers who are blind or visually impaired because there are, if you don't know them it is because you haven't investigated them in depth, just to name a few countries in which there are some math teachers is in the country of Costa Rica as well as in Spain, do not be one of those people or teachers who only say that a visually impaired student can not study mathematics because he is blind or visually impaired so you are in error, the visually impaired student can achieve it, just that you as a math teacher do not give those tools so that the student can overcome for example that you drive the training, If you think that there is only the Cranmer Abacus, the talking scientific calculator, the geoplane, the tangram, you are very wrong because there are many more, sometimes regular school teachers do not have the knowledge that there are some of these tools because they do not usually investigate it or because most of the time they have never worked with a student with visual impairment so regular school teachers say why training if we do not work with such students so in this project I invite you to promote training if you are a teacher of mathematics or other subject; [1]The person who is writing this book is a visually impaired person and she is showing you that she can be successful in a branch of mathematics.

<u>*Index*</u>

1

Prologue

I want to invite you as a math teacher to put yourself in the shoes of the visually impaired student so that he can learn better mathematics and can be encouraged to study a degree in mathematics, physics or engineering with your support could be achieved, only that the math teacher needs to have adequate preparation to receive a student with these characteristics because the teacher is not trained because most of them have no idea how a student with visual impairment can learn mathematics and a student with these characteristics is frustrated for various reasons because he usually faces many difficulties; If the student is blind, he/she usually uses Braille language and the teacher does not understand it, so the blind student cannot communicate with the teacher in a good way. This is one of the difficulties, but there are many more, and if the student is visually impaired, one of the difficulties is that he/she cannot see the blackboard in most of the occasions, even if he/she is in the chair in front of him/her, he/she cannot see it, so the teacher must have alternatives to teach this student, but the student must also do his/her part, the teacher is not going to solve the student's life 100% and neither can he/she pass the grade just because of his/her disability, that would be unethical, so the teacher as well as the student have to be a team in order to achieve the proposed objectives and sometimes what is prioritized is that the student has the passing grades to pass the year when this is a wrong perception since the student worries first about passing the year before learning, that is why the level of knowledge in several countries is very low.

I invite you to have this book at home and if you are a math teacher to be aware and take the initiative to go to courses or take training, if the math teacher really wants the visually impaired student you expect; support him because tomorrow you as a teacher could have a student with visual impairment and it would give you a helplessness because you will not know how to teach the student and this would be a great difficulty for you as a teacher as well as for the student and you will find yourself in a very serious problem.

As a visually impaired I invite you as a teacher to take trainings, courses, among other things. Because the visually impaired student has a lot of potential, but cannot develop it because he does not have the necessary tools, since the student also has the duty to look for them; As a lived experience it is very impotent not to be able to go at the same pace as the normo-visual students and even if you make an effort every day you do not understand some topics and if you ask the teacher he does not know how to help you because he is not prepared.

As the author of this book, I am trying to encourage the mathematics teaching community to take the challenge and prepare themselves because at any moment a visually impaired student may come to you and you will not be ready to explain the topics, think about this and you will come to the conclusion that you need to be trained for these situations.

Chapter 1: Calculating the perimeter and area of a regular pentagon.

Definition of regular pentagon

Regular pentagon.

A plane geometric figure that has five sides and five equal angles.

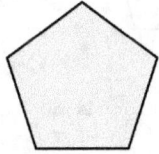

The sum of its internal angles is equal to 450 degrees and it has two diagonals.

To find how many diagonals a regular pentagon has, we do the following:

$$D= \frac{N(N-3)}{2}$$

Apothema of a regular pentagon.

It is the perpendicular distance from the center of the regular pentagon to the center of one of its sides.

The apothem of a regular pentagon is calculated as follows:

$$A= \frac{l}{2ta(B)}$$

The perimeter of a regular pentagon is calculated as follows:

$$P= 5 \times L$$

The area of a regular pentagon is calculated as follows:

$$A= \frac{Perimetro \times apotema}{2}$$

1.2 Exercises calculating the perimeter and area of a regular pentagon.

Example 1:

Calculate   the   perimeter of a regular pentagon
having side a=2cm

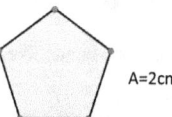

A=2cm

The perimeter of this regular pentagon is found as follows:

Using the perimeter formula

P= 5 × L

P= 5 × 2cm

P= 10cm

Answer: The perimeter is 10cm

Example 2:

Calculate the perimeter of a regular pentagon having side a=3cm.

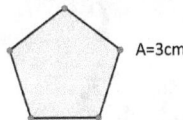 A=3cm

The perimeter of this regular pentagon is found as follows:

Using the perimeter formula

P= 5 × L

P= 5 × 3cm

P= 15cm

Answer: The perimeter is 15cm

Example 3:

Calculate the perimeter of a regular pentagon having side a=10cm.

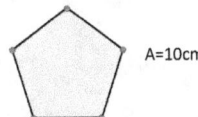 A=10cm

The perimeter of this regular pentagon is found as follows:

Using the perimeter formula

P= 5 × L

P= 5 × 10cm

P= 50cm

Answer: The perimeter is 50cm

Example 4:

Calculate the perimeter of a regular pentagon having side a=1m.

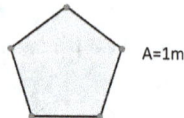

A=1m

The perimeter of this regular pentagon is found as follows:

Using the perimeter formula

P= 5 × L

P= 5 × 1m

P= 5m

Answer: The perimeter is 5m

Example 5:

Calculate the perimeter of a   **regular pentagon having side**
a=4m.

A=4m

The perimeter of this regular pentagon is found as follows:

Using the perimeter formula

P= 5 × L

P= 5 × 4m

P= 20m

Answer: The perimeter is 20cm

Example 6:

Find the side of a regular pentagon that has an apothem of 5cm.

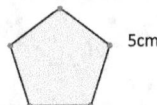
5cm

The side of this regular pentagon is found as follows:

Using the following formula

L= apothem × 2 tan T

The angle T is as follows

$$T = \frac{360 \ grados}{5 \times 2}$$

$$T = \frac{360 \ grados}{10}$$

T= 36 degrees

L= 5cm × 2tan (36) degrees.

L= 10 × tan (36) degrees

L= 7.26 cm

Answer: Side is 7.26cm

Example 7:

Find the side of a regular pentagon that has an apothem of 2m.

2m

The side of this regular pentagon is found as follows:

Using the following formula

L= apothem × 2tan T

Angle T is found as follows:

$$T= \frac{\frac{360 \; grados}{5 \; \times \; 2}}{}$$

$$T= \frac{\frac{360 \; grados}{10}}{}$$

T= 36 degrees.

Then the following is done

L= apothem × 2tan (36) degrees

L= 2m × 2 tan (36) degrees

L= 4m × tan (36) degrees

L= 2.90m

Answer: Side is 2.90m

Example 8:

Calculate the apothem of a regular pentagon with one of its sides measuring 4cm.

L=4cm

The apothem of this regular pentagon is found as follows:

Using the following formula

$$A= \frac{l}{2 \; \times \; tanT}$$

Angle T is found as follows:

$$T= \frac{360 \; grados}{5 \; \times \; 2}$$

$$T= \frac{360\ grados}{10}$$

T= 36 degrees.

$$A= \frac{4cm}{2\ \times\ tan(36)\ grados}$$

$$A= \frac{4cm}{1.45}$$

A= 2.75cm

Answer: The apothem is 2.75cm.

Example 9:

Calculate the apothem of a regular pentagon having a side of 3m.

L=3m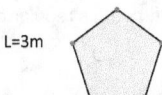

The apothem of this regular pentagon is found as follows:

Using the following formula

$$A= \frac{l}{2\ \times\ tanT}$$

Angle T is found as follows:

$$T= \frac{360\ grados}{5\ \times\ 2}$$

$$T= \frac{360\ grados}{10}$$

T= 36 degrees

$$A= \frac{3m}{2\ \times\ tan(36)\ grados}$$

$$A= \frac{3m}{1.45}$$

A= 2.06m

Answer: The apothem is 2.06m.

Example 10:

Calculate the area of a regular pentagon having a side of 10cm and an apothem of 5cm.

10cm

5cm

The area of this regular pentagon is found as follows:

Using the area formula

A= $\dfrac{perímetro \quad \times \quad apotema}{2}$

Before using this formula, the perimeter of this regular pentagon will be found.

P= 5 × L

P= 5 × 10cm

P= 50cm

A= $\dfrac{perímetro \quad \times \quad apotema}{2}$

A= $\dfrac{50 \times 5}{2}$

A= $\dfrac{250}{2}$

A= 125cm ^{2}

Answer: The area is 125cm ^{2}

Exercise 1

Calculate the perimeter and area of a regular pentagon having a side of 4cm and an apothem of 2cm.

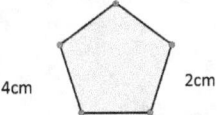

4cm   2cm

Solution:

First, the perimeter of this regular pentagon will be found as follows:

Using the perimeter formula

P= 5 × L

P= 5 × 4cm

P= 20cm

Then the area is found, using the area formula

$$A= \frac{perímetro \times apotema}{2}$$

$$A= \frac{20 \times 2}{2}$$

$$A= \frac{40}{2}$$

A= 20cm^2

Answer: The perimeter is 20cm and the area is 20cm^2

Exercise 2

Calculate the perimeter and area of a regular pentagon having a side of 6cm and an apothem of 4cm.

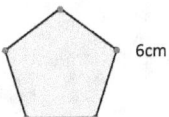

6cm

Solution:

First, the perimeter of this regular pentagon will be found as follows:

Using the perimeter formula

P= 5 × L

P= 5 × 6cm

P= 30cm

Then the area is found, using the area formula

$$A= \frac{perímetro \times apotema}{2}$$

$$A= \frac{30 \times 4}{2}$$

$$A= \frac{120}{2}$$

A= 60cm ^{2}

Answer: The perimeter is 30cm and the area is 60cm ^{2}

Exercise 3

Calculate the perimeter and area of a regular pentagon having a side of 3cm and an apothem of 1.5cm.

1.5cm

3cm

Solution:

First, the perimeter of this regular pentagon will be found as follows:

Using the perimeter formula

P= 5 × L

P= 5 × 3cm

P= 15cm

Then the area is found, using the formula for the area of a regular pentagon

$$A= \frac{perímetro \quad \times \quad apotema}{2}$$

$$A= \frac{15 \quad \times \quad 1.5}{2}$$

$$A= \frac{22.5}{2}$$

A= 11.25cm^2

Answer: The perimeter is 15cm and the area is 11.25cm^2

Exercise 4

Calculate the perimeter and area of a regular pentagon having an apothem of 20 cm.

20cm

Solution:

First, the side of this regular pentagon will be found as follows:

L= apothem × 2tanT

Angle T is found as follows:

$$T= \frac{360 \quad grados}{5 \quad \times \quad 2}$$

$$T= \frac{360 \quad grados}{10}$$

T= 36 degrees

L= 20cm × 2tan (36) degrees

L= 40 × tan (36) degrees

L= 29.06cm

The perimeter is found using the following formula:

P= 5 × L

P= 5 × 29.06

P= 145.3cm

Then the area is found, using the formula for the area of a regular pentagon.

$$A= \frac{perímetro \times apotema}{2}$$

$$A= \frac{145.3 \times 20}{2}$$

$$A= \frac{2906}{2}$$

A= 1453 cm ^{2}

Answer: The perimeter is 145.3cm and area is 1453 cm ^{2}

Exercise 5

Calculate the perimeter and area of a regular pentagon having a side of 9m.

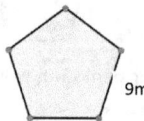

9m

Solution:

First, the apothem of this regular pentagon will be found as follows:

$$A= \frac{l}{2 \times tanT}$$

Angle T is found as follows:

$$T= \frac{360 \; grados}{5 \times 2}$$

$$T= \frac{360 \; grados}{10}$$

T= 36 degrees.

$$A= \frac{9}{2 \times tan(36) \; grados}$$

$$A= \frac{9}{1.45}$$

A= 6.20m

Find the perimeter using the perimeter formula

P= 5 × L

P= 5 × 9m

P= 45m

Then the area is found using the following formula:

$$A= \frac{perímetro \quad × \quad apotema}{2}$$

$$A= \frac{45 \quad × \quad 6.20}{2}$$

$$A= \frac{279}{2}$$

A= 139.5m ^{2}

Answer: The perimeter is 45m and the area is 139.5m ^{2}

1.3 Problems calculating the perimeter and area of a regular pentagon

Problem 1:

Calculate the perimeter of a regular pentagon having a side of 3cm and an apothem of 1.2cm.

3cm 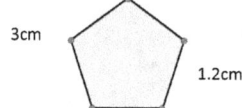 1.2cm

Solution:

This problem is solved as follows:

Using the perimeter formula

P= 5 × L

P= 5 × 3 cm

P= 15 cm

Answer: The perimeter is 15 cm

Problem 2:

If the area of a regular pentagon is 5m² and its apothem is 1.17m, how long is one of its sides?

5m 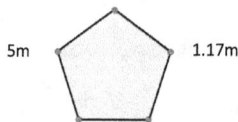 1.17m

Solution:

This problem is solved as follows:

Taking as a reference the formula for the area

$$A= \frac{5 \times l \times ap}{2}$$

Since it is known how much the area and apothem measure, then the values of the area and apothem are substituted into the area formula.

$$5= \frac{5 \ \times \ l \ \times \ 1.17}{2}$$

Then clear L

$$L= \frac{5 \times 2}{5 \ \times \ 1.17}$$

$$L= \frac{2}{1.17}$$

L= 1.71m

Answer: The side measures 1.71m

Problem 3:

Calculate the area of a regular pentagon having a perimeter of 100m and an apothem of 5m.

5m   P=100m

Solution:

How this problem is solved is as follows:

Using the area formula

$$A= \frac{perímetro \ \times \ ap}{2}$$

$$A= \frac{100 \ \times \ 5}{2}$$

$$A= \frac{500}{2}$$

A= 250m ^{2}

Answer: The area is 250m²

Problem 4:

Calculate the area of a regular pentagon having a perimeter of 50m and an apothem of 4m.

4m   P=50m

Solution:

This problem is solved as follows:

Using the area formula

$$A= \frac{perímetro \times ap}{2}$$

$$A= \frac{50 \times 4}{2}$$

$$A= \frac{200}{2}$$

A= 100m ^{2}

Answer: The area is 100m ^{2}

Problem 5:

Prove that the formula for the area of a regular pentagon is as follows:

$$A= \frac{5 \times l \times ap}{2}$$

Solution:

As shown, the formula for the area is as follows:

First draw a regular pentagon with the apothem that will form 5 triangles inside the regular pentagon.

Looking at the figure you can see that the area of the regular pentagon is 5 times the area of one of these triangles.

Then the following is done

The area of one of these triangles inside the regular pentagon is calculated.

Using the following formula $A= \frac{b \times h}{2}$

The base of one of the triangles taken coincides with a side of the regular pentagon.

The height of this triangle coincides with the apothem of the regular pentagon.

This leaves **A=** $\dfrac{l \times ap}{2}$

Then multiply by 5 to obtain the area of the regular pentagon.

Then it remains

A= $\dfrac{5 \times l \times ap}{2}$

Chapter 2: Calculating the perimeter and area of a regular hexagon.

Definition of regular hexagon

Regular hexagon.

It is a polygon that has six equal sides and six equal angles.

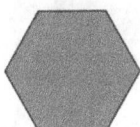

By joining the center of the regular hexagon with its vertices, six equilateral triangles are formed inside the regular hexagon.

The sum of its internal angles is (6 - 2), so 180 degrees= 720 degrees.

The value of an internal angle of a regular hexagon

$$\frac{720 \ grados}{6} = 120 \ \textbf{degrees.}$$

The central angle of a regular hexagon measures 60 degrees.

It is found as follows:

$$\frac{360 \ grados}{6} = 60 \ \textbf{degrees.}$$

The regular hexagon has 9 diagonals and they are as follows:

$$6(\frac{6-3}{2})$$

$$6(\frac{3}{2})$$

$$\frac{18}{2} = 9$$

Therefore, the regular hexagon has 9 diagonals.

Apothema of the regular hexagon.

It is the distance from the center to one of its sides.

The apothem of a regular hexagon is calculated as follows:

L= r

The perimeter of a regular hexagon is calculated as follows:

$$P= 6 \times L$$

The area of a regular hexagon is calculated as follows:

$$A= \frac{p \times ap}{2}$$

2.2 Exercises calculating the perimeter and area of a regular hexagon

Example 1:

Calculate the perimeter of a regular hexagon having a side of 2cm.

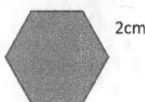
2cm

The perimeter of this regular hexagon is found as follows:

Using the perimeter formula

P= 6 × L

P= 6 × 2cm

P= 12cm

Answer: The perimeter is 12cm

Example 2:

Calculate the perimeter of a regular hexagon having a side of 5cm.

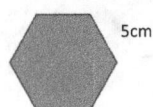
5cm

The perimeter of this regular hexagon is found as follows:

Using the perimeter formula

P= 6 × L

P= 6 × 5cm

P= 30 cm

Answer: The perimeter is 30cm

Example 3:

Calculate the perimeter of   **a regular hexagon having a side of 3cm.**

3cm

The perimeter of this regular hexagon is as follows:

Using the perimeter formula

P= 6 × L

P= 6 × 3 cm

P= 18 cm

Answer: The perimeter is 18cm

Example 4:

Calculate the perimeter   **of a regular hexagon having a side of 10m.**

10m

The perimeter of this regular hexagon is as follows:

Using the perimeter formula

P= 6 × L

P= 6 × 10m

P= 60m

Answer: The perimeter is 60 m

Example 5:

Calculate the side of a   regular   hexagon   having   an
apothem of 10 m.

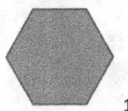

10m

The side of this regular hexagon is found as follows:

Inside the regular hexagon you will have 6 equilateral triangles drawn, you will have to extract one of the right triangles and then use the Pythagorean theorem.

Since there are 6 equilateral triangles inside the regular hexagon, the sides of the equilateral triangles are equal so the unknown sides are given x.

Using the Pythagorean theorem.

$2y^2 = 10^2 + y^2$

Y was added to facilitate the equation

$4y^2 = 100 + y^2$

$4y^2 - y^2 = 100$

$3y^2 = 100$

It is then divided and **remains** $\dfrac{3y^2}{3} = \dfrac{100}{3}$

$Y^2 = \dfrac{100}{3}$

The square root is taken and **the result is** $\sqrt{y^2} = \sqrt{\dfrac{100}{3}}$

Y= 5.77m

Therefore, the side of this regular hexagon is 5.77m.

Answer: Side is 5.77m

Example 6:

Calculate the side of a **regular hexagon having an apothem of 15m.**

15m

How the side of this regular hexagon is found is as follows:

First draw the regular hexagon with its apothem of 15 m and inside the hexagon will have 6 equilateral triangles in which one of them will be extracted and then use the Pythagorean theorem. We will put X to the unknown sides.

Using the Pythagorean theorem, it looks like this

$2y^2 = 15^2 + y^2$

$4y^2 = 225 + y^2$

$4y^2 - y^2 = 225$

$3y^2 = 225$

Then divide by 3 and **you get** $\dfrac{3y^2}{3} = \dfrac{225}{3}$

Y² = 75

The square root is taken and **the result is** $\sqrt{y^2} = \sqrt{75}$

Y= 8.66m

Therefore, the side of this regular hexagon is 8.66m.

Answer: Side is 8.66m

Example 7:

Calculate the apothem of a   **regular hexagon having a side of**
4cm.

4cm

The apothem of this regular hexagon is found as follows:

Using the following formula in order to find the angle.

$$\frac{360 \;\; grados}{6 \;\; \times \;\; 2}$$

$$\frac{360 \;\; grados}{12} = 30 \;\; \textbf{degrees}$$

Then the following formula is used to find the apothem.

A= $\dfrac{l}{2tanT}$

A= $\dfrac{4}{2 \;\; \times \;\; tan(30) \;\; grados}$

A= $\dfrac{4}{1.14}$

A= 3.50cm

Answer: The apothem is 3.50cm.

Example 8:

Calculate the apothem of a regular hexagon having a side of 3cm.

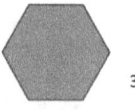

3cm

The apothem of this regular hexagon is as follows:

First the following formula will be used in order to find the angles.

$$\frac{360 \; grados}{6 \times 2}$$

$$\frac{360 \; grados}{12} = \textbf{30 degrees}$$

Then the following formula is used to find the apothem of this regular hexagon.

$$\textbf{A=} \frac{l}{2 \times tanT}$$

$$\textbf{A=} \frac{3}{2 \times ta(30)}$$

$$\textbf{A=} \frac{3}{1.14}$$

A= 2.63cm

Answer: The apothem is 2.63cm

Example 9:

Calculate the area of a regular hexagon having a side of 12 cm and an apothem of 5 cm.

The area of this regular hexagon is found as follows:

First, the perimeter of this regular hexagon will be found.

Using the perimeter formula

P= 6 × L

P= 6 × 12cm

P= 72 cm

Then the area is calculated using the area formula

$$A= \frac{p \times ap}{2}$$

$$A= \frac{72 \times 5}{2}$$

$$A= \frac{360}{2}$$

A= 180 cm ^{2}

Answer: The area is 180 cm ^{2}

Exercises

Exercise 1

Calculate the perimeter and area of a regular hexagon having a side of 2cm and an apothem of 1.5cm.

1.5cm   2cm

Solution:

First, the perimeter of this regular hexagon will be found as follows:

P= 6 × L

P= 6 × 2 cm

P= 12 cm

Then the area of this regular hexagon is found as follows:

$$A= \frac{p \times ap}{2}$$

$$A= \frac{12 \times 1.5}{2}$$

$$A= \frac{18}{2}$$

A=9cm ^{2}

Answer: The perimeter is 12 cm and the area is 12 cm ^{2}

Exercise 2

Calculate the perimeter and area of a regular hexagon having a side of 10 cm and an apothem of 3 cm.

3cm  10cm

Solution:

First, the perimeter of this regular hexagon will be found as follows:

P= 6 × L

P= 6 × 10 cm

P= 60 cm

Then the area of this regular hexagon is found as follows:

$$A= \frac{p \times ap}{2}$$

$$A= \frac{60 \times 3}{2}$$

$$A= \frac{180}{2}$$

A= 90 cm ^{2}

Answer: The perimeter is 60cm and the area is 90 cm ^{2}

Exercise 3

Calculate the perimeter and area of a regular hexagon having a side of 20 cm and an apothem of 10 cm.

10cm   20cm

Solution:

First, the perimeter of this regular hexagon will be found as follows:

P= 6 × L

P= 6 × 20cm

P= 120cm

Then the area of this regular hexagon is found as follows:

$$A= \frac{p \times ap}{2}$$

$$A= \frac{120 \times 10}{2}$$

$$A= \frac{1200}{2}$$

A= 600cm ^{2}

Answer: The perimeter is 120cm and the area is 600cm ^{2}

Exercise 4

Calculate the perimeter and   area of a regular hexagon having a side of 6m.

6m

Solution:

First, the following formula will be used to find the angle T

$$T= \frac{360 \ grados}{6 \ \times \ 2}$$

$$T= \frac{360 \ grados}{12}$$

T= 39 degrees

The following formula is used to find the apothem

$$A= \frac{l}{2 \ \times \ tanT}$$

$$A= \frac{6}{2 \ \times \ tan(30) \ grados}$$

$$A= \frac{6}{1.14}$$

A= 5.26m

Then the perimeter of this regular hexagon will be found as follows:

P= 6 × L

P= 36m

Then the area of this regular hexagon is found as follows:

$$A= \frac{p \ \times \ ap}{2}$$

$$A= \frac{36 \ \times \ 5.26}{2}$$

$$A= \frac{189.36}{2}$$

A= 94.68m ^{2}

Answer: The perimeter is 36m and the area is 94.68m ^{2}

Exercise 5

Calculate the perimeter and area of a regular hexagon having an apothem of 50 m.

Solution:

First, the following formula will be used to find the angle T

$$T= \frac{360 \quad grados}{6 \times 2}$$

$$T= \frac{360}{12}$$

T= 30 degrees

Then the following formula is used to find the side of this regular hexagon.

L= ap × 2tanT

L= 50 × 2 × tan(30) degrees

L= 50 × 1.14

L= 57m

The perimeter is found as follows:

P= 6 × L

P= 6 × 57m

P= 342 m

Then the area is as follows:

$$A= \frac{p \times ap}{2}$$

$$A= \frac{342 \times 50}{2}$$

$$A= \frac{17100}{2}$$

A= 8550m ^{2}

Answer: The perimeter is 342m and the area is 8550m ^{2}

2.3 Problems of calculating the perimeter and area of a regular hexagon.

Problem 1:

Calculate the perimeter and area of a regular hexagon having a side of 2.5m.

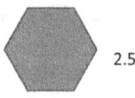

2.5m

Solution:

First, the perimeter of this regular hexagon will be found as follows:

P= 6 × L

P= 6 × 2.5m

P= 15m

Then the area is found with the following formula

$$A= \frac{p \ \times \ ap}{2}$$

This formula cannot yet be applied because the apothem of this regular hexagon is missing.

The apothem of this regular hexagon is as follows.

First, the angle T

$$T= \frac{360 \ grados}{6 \ \times \ 2}$$

$$T= \frac{360 \ grados}{12}$$

T= 30 degrees

Then the following formula is used to find the apothem

$$A= \frac{l}{2tanT}$$

$$A= \frac{2.5}{2 \ \times \ tan(30) \ grados}$$

$$A= \frac{2.5}{1.14}$$

A= 2.19m

Now you can use the formula for area

$$A= \frac{p \times ap}{2}$$

$$A= \frac{15 \times 2.19}{2}$$

$$A= \frac{34.35}{2}$$

A= 17.17m ^{2}

Answer: The perimeter is 15m and the area is 17.17m.

Problem 2:

Calculate the perimeter and area of a regular hexagon having a side of 10m and an apothem of 4m.

10m

4m

Solution:

First, the perimeter will be found as follows:

P= 6 × L

P= 6 × 10m

P= 60m

Then there is the area

$$A= \frac{p \times ap}{2}$$

$$A= \frac{60 \times 4}{2}$$

$$A= \frac{240}{2}$$

A= 120m ^{2}

Answer: The perimeter is 60m and the area is 120m ^{2}

Chapter 3: Calculating the area and volume of a cube.

Definition of cube.

Cube

It is a polyhedron formed by six equal squares.

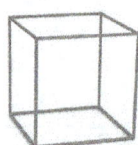

The cube has the following properties:

- Six faces
- Eight vertices
- Twelve edges

The area of a cube is calculated as follows:

$$A = 4 \times a^2$$

It can also be calculated as follows:

$$A = 6 \times a^2$$

The volume of a cube is calculated as follows:

$$V = a^3$$

3.2 Exercises calculate the area and volume of a cube.

Example 1:

Calculate the area of a cube that has a side of 2cm.

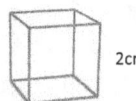 2cm

The area of this cube is found as follows:

Using the area formula

A= 6 × L ^{2}

A= 6 × 2 cm ^{2}

A= 6 × 4 cm ^{2}

A= 24 cm ^{2}

Answer: The area is 24 cm ^{2}

Example 2:

Calculate the area of a cube that has a side of 5cm.

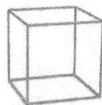 5cm

How the area of this cube is found is as follows:

A= 6 × L ^{2}

A= 6 × 5cm ^{2}

A= 6 × 25cm ^{2}

A= 150cm ^{2}

Answer: The area is 150cm ^{2}

Example 3:

Calculate the area of a cube having a side of 4m.

 4m

The area of this cube is found as follows:

A= 6 × L ^{2}

A= 6 × 4m ^{2}

A= 6 × 16m ^{2}

A= 96m ^{2}

Answer: The area is 96m ^{2}

Example 4:

Calculate the area of a cube having a side of 3m.

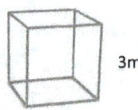 3m

The area of this cube is found as follows:

A= 6 × L ^{2}

A= 6 × 3m ^{2}

A= 6 × 9m ^{2}

A= 54m ^{2}

Answer: The area is 54m^2

Example 5:

Calculate the volume of a cube having a side of 2 cm

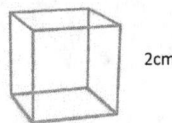
2cm

As found the volume of this cube is as follows:

$V = L^3$

$V = 2cm^3$

$V = 8\ cm^3$

Answer: The volume is 8 cm³

Example 6:

Calculate the volume of a cube having a side of 5cm.

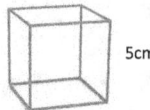
5cm

The volume of this cube is found as follows:

$V = L^3$

$V = 5cm^3$

$V = 125cm^3$

Answer: The volume is 125cm³

Example 7:

Calculate the volume of a cube having a side of 3cm.

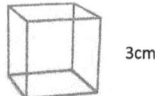

3cm

As found the volume of this cube is as follows:

V= L ^{3}

V= 3cm ^{3}

V= 27cm ^{3}

Answer: The volume is 27cm ^{3}

Example 8:

Calculate the volume of a cube having a side of 10m.

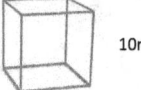

10m

As found the volume of this cube is as follows:

V= L ^{3}

V= 10m ^{3}

V= 1000m ^{3}

Answer: The volume is 1000m ^{3}

Example 9:

Calculate the volume of a cube having a side of 6m.

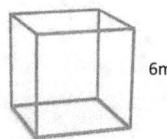

6m

As found the volume of this cube is as follows:

V= L ^{3}

V= 6m ^{3}

V= 216m^{3+}

Answer: The volume is 216m ^{3}

Exercises

Exercise 1

Calculate the area and volume of a cube having a side of 1cm.

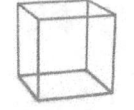

1cm

Solution:

First, the area of this cube is found as follows:

A= 6 × L ^{2}

A= 6 × 1cm ^{2}

A= 6cm ^{2}

Then the volume of this cube is found as follows:

$V= L^3$

$V= 1cm^3$

$V= 1cm^3$

Answer: The area is $6cm^2$ and the volume is $1cm^3$

Exercise 2

Calculate the area and volume of a cube that has a side of 2.5cm.

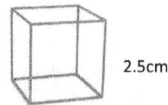
2.5cm

Solution:

First, there is the area

$A= 6 \times L^2$

$A= 6 \times 1.5cm^2$

$A= 6 \times 6.25cm^2$

$A= 37.5cm^2$

Then there is the volume

$V= L^3$

$V= 2.5cm^3$

$V= 15.625cm^3$

Answer: The area is $37.5cm^2$ and the volume is $15.625cm^3$

Exercise 3

Calculate the area and volume of a cube having a side of 4cm.

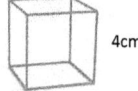
4cm

Solution:

First, the area of this cube will be found as follows:

$A = 6 \times L^2$

$A = 6 \times 4 \text{ cm}^2$

$A = 6 \times 16 \text{ cm}^2$

$A = 96 \text{ cm}^2$

Then there is the volume

$V = L^3$

$V = 4 \text{ cm}^3$

$V = 64 \text{ cm}^3$

Answer: The area is 96 cm² and the volume is 64 cm³

Exercise 4

Calculate the area and volume of a cube having a side of 7m.

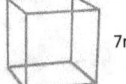 7m

Solution:

First, the area of this cube is found as follows:

$A = 6 \times L^2$

$A = 6 \times 7m^2$

$A = 6 \times 49m^2$

$A = 294m^2$

Then there is the volume

$V = L^3$

$V = 7m^3$

$V = 343m^3$

Answer: The area is 294m² and the volume is 343m ³

Exercise 5

Calculate the area and volume of a cube having a side of 8m.

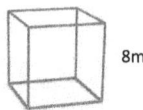 8m

Solution:

First, the area of this cube is found as follows:

A= 6 × L ²

A= 6 × 8m ²

A= 6 × 64m

A= 384m ²

Then there is the volume

V= L ³

V= 8m ³

V= 512m ³

Answer: The area is 384m² and the volume is 512m ³

Chapter 4: Calculating the area and volume of a sphere.

Definition of sphere.

Sphere

It is determined as the region of space that would be found inside a spherical surface.

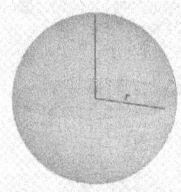

Elements of the sphere

- Center.
- Radio.
- Rope.
- Diameter.
- Poles.

Definition of the elements of a sphere

Center:

It is the lower point equidistant from any point on the surface of a sphere.

Radio:

It is the distance from the center to a point on the surface of the sphere.

Rope:

It is the segment that can join 2 points on the surface of a sphere.

Diameter:

It is the string passing through the center of the sphere.

Poles:

These are the points of the axis of rotation that lie on the spherical surface.

Calculating the radius of a sphere.

The radius of the sphere is calculated as follows

Knowing the distance of a plane that cuts the sphere and the radius of the section by applying the Pythagorean theorem.

$$R^2 = d^2 + r^2$$

This leaves **R=** $\sqrt{d^2 + r^2}$

The area of a sphere is calculated as follows:

$$A = 4 \times pi \times r^2$$

The volume of a sphere is calculated as follows:

$$V = \frac{4}{3}pi \times r^3$$

4.2 Exercises calculate the area and volume of a sphere.

Example 1:

Calculate the area of a sphere having a radius of 1 cm.

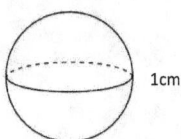 1cm

How the area of this sphere is found is as follows:

A= 4 × pi × r ^{2}

A= 4 × pi × 1 ^{2}

A= 4 × 3.14 × 1

A= 12.56cm ^{2}

Answer: The area is 12.56cm ^{2}

Example 2:

Calculate the area of a sphere having a radius of 2cm.

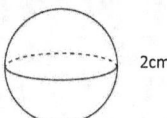 2cm

How to calculate the area of this sphere is as follows:

A= 4 × pi × r ^{2}

A= 4 × pi × 2 ^{2}

A= 4 × 3.14 × 4

A= 50.24cm ^{2}

Answer: The area is 50.24cm ^{2}

Example 3:

Calculate the area of a sphere having a radius of 10 cm.

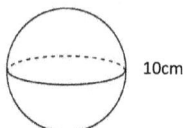 10cm

The area of this sphere is found as follows:

A= 4 × pi × r ^{2}

A= 4 × pi × 10 ^{2}

A= 4 × 3.14 × 100

A= 1256 cm ^{2}

Answer: The area is 1256 cm ^{2}

Example 4:

Calculate the area of a sphere having a radius of 5cm.

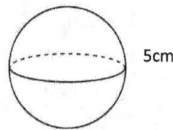 5cm

I low the area of this sphere is found is as follows:

A= 4 × pi × r ^{2}

A= 4 × pi × 5 ^{2}

A= 4 × 3.14 × 25

A= 314 cm ^{2}

Answer: The area is 314 cm ^{2}

Example 5:

Calculate the area of a sphere having a radius of 3m.

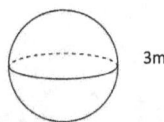 3m

How the area of this sphere is found is as follows:

A= 4 × pi × r ^{2}

A= 4 × pi × 3 ^{2}

A= 4 × 3.14 × 9

A= 113.04m ^{2}

Answer: The area is 113.04m ^{2}

Example 6:

Find the radius of a sphere having an area of 100cm ^{2}

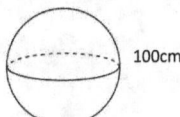 100cm

How the radius of this sphere is found is as follows:

A= 4 × pi × r ^{2}

100cm^2 = 4 × pi × r ^{2}

100cm^2 = 4 × 3.14 × r ^{2}

100cm^2 = 12.56 × r ^{2}

$$\frac{100cm^2}{12.56} = r^2$$

7.96 cm^2 = r ^{2}

Then we take the square root and **we get** $\sqrt{7.96cm^2} = \sqrt{r^2}$

2.82cm= r

Answer: The radius is 2.82cm.

Example 7:

Find the radius of a sphere having a volume of 1000cm ^{3}

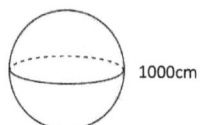

1000cm

How the radius of this sphere is found is as follows:

$$V = \frac{4 \times pi \times r^3}{3}$$

$$1000cm^3 = \frac{4 \times 3.14 \times r^3}{3}$$

$$1000cm^3 = \frac{12.56 \times r^3}{3}$$

$$3 \times 1000cm^3 = 12.56 \times r^3$$

$$3000cm^3 = 12.56 \times r^3$$

$$\frac{3000cm^3}{12.56} = r^3$$

$$238.85cm^3 = r^3$$

Then the cube root is taken out and **the result is** $\sqrt[3]{238.85cm^3} = \sqrt[3]{r^3}$

$$6.20cm = r$$

Answer: The radius is 6.20cm.

Example 8:

Calculate the volume of a sphere having a radius of 5cm.

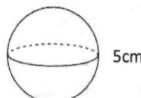

5cm

As found the volume of this sphere is as follows:

$$V= \frac{4 \times pi \times r^3}{3}$$

$$V= \frac{4 \times 3.14 \times 5^3}{3}$$

$$V= \frac{12.56 \times 125}{3}$$

$$V= \frac{1570}{3}$$

$$V= 523.33cm^3$$

Answer: The volume is 523.33cm³

Example 9:

Calculate the volume of a sphere having a radius of 2cm.

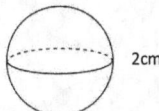 2cm

As found the volume of this sphere is as follows:

$$V= \frac{4 \times pi \times r^3}{3}$$

$$V= \frac{4 \times 3.14 \times 2^3}{3}$$

$$V= \frac{12.56 \times 8}{3}$$

$$V= \frac{100.48}{3}$$

$$V= 33.49cm^3$$

Answer: Volume is 33.49cm³

Example 10:

Calculate the volume of a sphere having a radius of 1 cm.

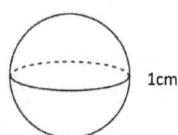 1cm

As found the volume of this sphere is as follows:

$$V= \frac{4 \times pi \times r^3}{3}$$

$$V= \frac{4 \times 3.14 \times 1^3}{3}$$

$$V= \frac{12.56 \times 1}{3}$$

$$V= \frac{12.56}{3}$$

$V= 4.18cm^3$

Answer: The volume is 4.18cm ^{3}

Exercises

Exercise 1

Calculate the area and volume of a sphere having a radius of 3cm.

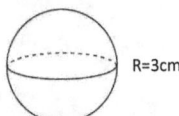 R=3cm

Solution:

First, the area of this sphere will be found as follows:

A= 4 × pi × r ^{2}

A= 4 × 3.14 × 3 ^{2}

A= 12.56 × 9

A= 113.04cm ^{2}

Then there is the volume

$$V= \frac{4 \times pi \times r^3}{3}$$

$$V= \frac{4 \times 3.14 \times 3^3}{3}$$

$$V= \frac{12.56 \times 9}{3}$$

$$V= \frac{339.12}{3}$$

V= 113.04cm ^{3}

Exercise 2

Calculate the area and volume of a sphere having a radius of 12cm.

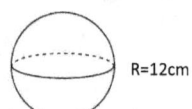 R=12cm

Solution:

First, the area of this sphere will be found as follows:

A= 4 × pi × r ²

A= 4 × pi × 12 ²

A= 4 × 3.14 × 144

A= 12,56 × 144

A= 1808.64cm ²

Then there is the volume

$$V= \frac{4 \times pi \times r^3}{3}$$

$$V= \frac{4 \times 3.14 \times 12^3}{3}$$

$$V= \frac{4 \times 3.14 \times 1728}{3}$$

$$V= \frac{12.56 \times 1728}{3}$$

$$V = \frac{21703.68}{3}$$

V= 7234.56cm ^{3}

Answer: The area is 1808.64cm^2 and the volume is 7234.56cm ^{3}

Chapter 5: Calculating the area and volume of a cylinder.

Definition of cylinder

Cylinder

It is a geometric body that is engendered by a rectangle that is rotated around one of its sides.

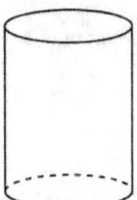

Elements of a cylinder

- Axis
- Basis
- Height
- Generatrix.

Definitions of the elements of a cylinder

Axis

It is determined as the fixed side around which the rectangle rotates.

Basis

They are the circles that generate the sides perpendicular to the axis.

Height

It is the distance between the two bases.

Generatrix

It is the side opposite the axis and is the side that will generate the cylinder.

The total area of a cylinder is calculated as follows:

$$A= 2 \times pi \times r \times h \times (h + r)$$

The area of the cylinder is calculated as follows:

A= 2 × pi × r × h + 2 × pi × r²

The volume of a cylinder is calculated as follows:

V= pi × r² × h

5.2 Exercises calculate area and volume of a cylinder.

Example 1:

Calculate the area of a cylinder having a radius of 2cm and a height of 5cm.

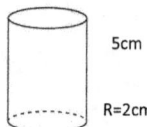
5cm

R=2cm

As found the area of this cylinder is as follows:

A= 2 × pi × r × h + 2 × pi × r ²

A= 2 × 3.14 × 2 × 5 + 2 × 3.14 × 2 ²

A= 6.28 × 2 × 5 + 2 × 3.14 × 4

A= 62.8 + 25.12

A= 87.92cm ²

Answer: The area is 87.92cm ²

Example 2:

Calculate the area of a cylinder having a radius of 2cm and a height of 1cm.

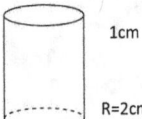

1cm

R=2cm

As found the area of this cylinder is as follows:

A= 2 × pi × r × h + 2 × pi × r ^{2}

A= 2 × 3.14 × 2 × 1 + 2 × 3.14 × 2 ^{2}

A= 6.28 × 2 × 1+ 6.28 × 4

A= 12.56 + 25.12

A= 37.68cm ^{2}

Answer: The area is 37.68cm ^{2}

Example 3:

Calculate the area of a cylinder having a radius of 2cm and a height of 1cm.

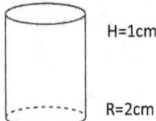

H=1cm

R=2cm

As found the area of this cylinder is as follows:

A= 2 × pi × r × h + 2 × pi × r ^{2}

A= 2 × 3.14 × 2 × 1+ 2 × 3.14 × 2 ^{2}

A= 6.28 × 2 × 1 + 6.28 × 4

A= 12.56 + 25.12

A= 37.68cm ^{2}

Answer: The area is 37.68cm ^{2}

Example 4:

Calculate the area of a cylinder having radius 5cm and height 3cm.

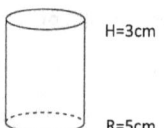

H=3cm

R=5cm

As found the area of this cylinder is as follows:

A= 2 × pi × r × h + 2 × pi × r ^{2}

A= 2 × 3.14 × 5 × 3 + 2 × 3.14 × 5 ^{2}

A= 6.28 × 5 × 3 + 6.28 × 25

A= 94.2 + 157

A= 251.2cm ^{2}

Answer: The area is 251.2cm ^{2}

Example 5:

Calculate the area of a cylinder having a radius of 10m and a height of 4m.

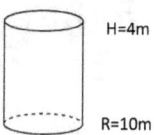

H=4m

R=10m

As found the area of this cylinder is as follows:

A= 2 × pi × r × h + 2 × pi × r ^{2}

A= 2 × 3.14 ×10 × 4 + 2 × 3.14 × 10 ^{2}

A= 6.28 × 10 × 4 + 6.28 × 100

A= 251.2 + 628

A= 779.2m ^{2}

Answer: The area is 779.2m ^{2}

Example 6:

Calculate the area of a cylinder having a radius of 4m and a height of 3m.

PA

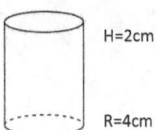

H=3m

R=4m

As found the area of this cylinder is as follows:

A= 2 × pi × r × h + 2 × pi × r²

A= 2 × 3.14 × 4 × 3 + 2 × 3.14 × 4 ²

A= 6.28 × 4 × 3 + 6.28 × 16

A= 75.36 + 100.48

A= 175.84m ²

Answer: The area is 175.84m²

Example 7:

Calculate the total area of a cylinder having a radius of 4cm and a height of 2cm.

H=2cm

R=4cm

The total area of this cylinder is found as follows:

A= 2 × pi × r × (h + r)

A= 2 × 3.14 × 4 × (2 + 4)

A= 2 × 3.14 × 4 × (6)

A= 6.28 × 4 × (6)

A= 150.72cm ²

Answer: The area is 150.72cm ²

Example 8:

Calculate the total area of a cylinder having a radius of 10cm and a height of 5cm.

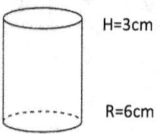

H=5cm

R=10cm

The total area of this cylinder is found as follows:

A= 2 × pi × r × (h + r)

A= 2 × 3.14 × 10 × (5+10)

A= 6.28 × 10 × (15)

A= 942cm ^{2}

Answer: The area is 942cm ^{2}

Example 9:

Calculate the total area of a cylinder having a radius of 6cm and a height of 3cm.

H=3cm

R=6cm

The total area of this cylinder is found as follows:

A= 2 × pi × r × (h + r)

A= 2 × 3.14 × 6 × (3 + 6)

A= 6.28 × 6 × (9)

A= 339.12cm ^{2}

Answer: The area is 339.12cm ^{2}

Example 10:

Calculate the total area of a cylinder having a radius of 12 cm and a height of 8 cm.

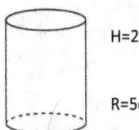

H=8cm

R=12cm

The total area of this cylinder is found as follows:

A= 2 × pi × r × (h + r)

A= 2 × 3.14 × 12 × (8 + 12)

A= 6.28 × 12 × (20)

A= 1507.2cm ^{2}

Answer: The area is 1507.2cm ^{2}

Example 11:

Find the volume of a cylinder having a radius of 5cm and a height of 2cm.

H=2

R=5c

The volume of this cylinder is found as follows:

V= pi × r^2 × h

V= 3.14 × 5^2 × 2

V= 3.14 × 25 × 2

V= 157 cm ^{3}

Answer: The volume is 157 cm ^{3}

Example 12:

Find the volume of a cylinder having a radius of 10cm and a height of 4cm.

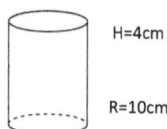

H=4cm

R=10cm

The volume of this cylinder is found as follows:

V= pi × r^2 × h

V= 3.14 × 10^2 × 4

V= 3.14 × 100 × 4

V= 1256 cm ^{3}

Answer: The volume is 1256 cm ^{3}

Example 13:

Find the volume of a cylinder having a radius of 2cm and a height of 6cm.

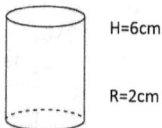

H=6cm

R=2cm

The volume of this cylinder is found as follows:

V= pi × r^2 × h

V= 3.14 × 2^2 × 6

V= 3.14 × 4 × 6

V= 75.36cm ^{3}

Answer: The volume is 75.36cm ^{3}

Example 14:

Find the volume of a cylinder having a radius of 3m and a height of 7m.

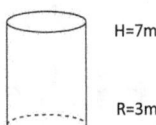

The volume of this cylinder is found as follows:

V= pi × r² × h

V= 3.14 × 3² × 7

V= 3.14 × 9 × 7

V= 197.82m ³

Answer: The volume is 197.82m ³

Exercises

Exercise 1

Calculate the area and volume of a cylinder having a radius of 8cm and a height of 3cm.

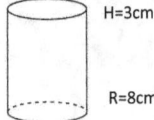

Solution:

First the area will be found as follows:

A= 2 × pi × r × h + 2 × pi × r ²

Then it remains

A= 2 × 3.14 × 8 × 3 + 2 × 3.14 × 8 ²

A= 6.28 × 8 × 3 + 6.28 × 64

A= 150.72 + 401.92

A= 552.64cm²

The volume is found as follows:

V= pi × r² × h

Then it remains

V= 3.14 × 8² × 3

V= 3.14 × 64 × 3

V= 602.88cm ³

Answer: The area is 552.64cm² and the volume is 602.88cm ³

Exercise 2

Calculate the area and volume of a cylinder having a radius of 10m and a height of 30m.

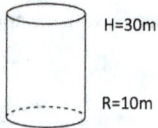
H=30m

R=10m

Solution:

First, the area of this cylinder will be found as follows:

A= 2 × pi × r × h + 2 × pi × r ²

Then it remains

A= 2 × 3.14 × 10 × 30 + 2 × 3.14 × 10 ²

A= 6.28 × 10 × 30 + 6.28 × 100

A= 21884 + 528

A= 2512m²

Then you will find the volume

V= pi × r² × h

Then it remains

V= 3.14 × 10² × 30

V= 3.14 × 100 × 30

V= 9420m ^{3}

Answer: The area is 2512m^2 and the volume is 9420m ^{3}

Chapter 6: Solving an obtuse-angled triangle

Definition of obtuse triangle

Obtuse-angled triangle.

A triangle with one obtuse angle and two acute angles.

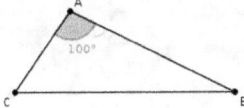

The perimeter of an obtuse triangle is calculated as follows:

$$P = a + b + c$$

The area of an obtuse triangle.

$$A = \frac{b \times h}{2}$$

6.2 Exercises solving an obtuse triangle.

Exercise 1

Solve an obtuse triangle having side b=10 cm, side a, side c, angle A=30 degrees, angle B=100 degrees and angle c.

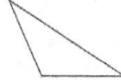

Solution:

How to solve this exercise is as follows:

The law of sines is used.

The sum of its 3 interior angles **= 180 degrees.**

It then follows that A + B + C= **180 degrees**

30 degrees + 100 degrees + C= 180 degrees

C= 180 degrees - 100 degrees - 30 degrees

C= 50 degrees

Angle C = 50 degrees.

Then there is the side a

$$\frac{b}{sen\ B} = \frac{a}{sen\ A}$$

$$\frac{10}{sen\ 100\ grados} = \frac{a}{sen\ 30\ grados}$$

It remains **that** $\dfrac{10}{0.98} = \dfrac{a}{0.5}$

$$\frac{10\ \times\ 0.5}{0.98} = a$$

$$\frac{5}{0.98} = a$$

5.10cm= a

Side a= 5.10cm

Then side c is found as follows:

Using the law of sines.

Then **it remains** $\dfrac{b}{sen\ B} = \dfrac{c}{sen\ C}$

$$\frac{10}{sen\ 100\ grados} = \frac{c}{sen\ 50\ grados}$$

Then it remains **that** $\dfrac{10}{0.98} = \dfrac{c}{0.76}$

$$\frac{10\ \times\ 0.76}{0.98} = c$$

$$\frac{7.6}{0.98} = c$$

7.75cm=c

Answer: Side a is 5.10cm, side b is 10 cm, side c is 7.75cm and angle c is 50 degrees.

Exercise 2

Solve an obtuse triangle having side a=20 cm, side b, side c, angle A=120 degrees, angle B=40 degrees and angle C.

Solution:

Using the law of sines.

First find the missing angle C in the obtuse triangle as follows:

Knowing that in any triangle the sum of its 3 interior angles=180 degrees

A + B + C = 180 degrees

Then it remains that **120 degrees + 40 degrees + C=** 180 degrees

C= 180 degrees -120 degrees - 40 degrees

C= 20 degrees

Angle C= 20 degrees

Then there is the side b

$$\frac{a}{sen\ A} = \frac{b}{sen\ B}$$

$$\frac{20}{sen\ 120\ grados} = \frac{b}{sen\ 40\ grados}$$

Then **it remains** $\frac{20}{0.86} = \frac{B}{0.64}$

$$\frac{20\ \times\ 0.64}{0.86} = b$$

$$\frac{12.8}{0.86} = b$$

14.88cm= b

The side c will be found using the law of sines

$$\frac{a}{sen\ A} = \frac{c}{sen\ C}$$

$$\frac{20}{sen\ 120\ grados} = \frac{c}{sen\ 20\ grados}$$

Then it remains $\dfrac{20}{0.86} = \dfrac{c}{0.34}$

$\dfrac{20 \times 0.34}{0.86} = c$

$\dfrac{6.8}{0.86} = c$

7.90cm= c

Answer: Side b is 14.88cm, side c is 7.90cm and angle c is 20 degrees.

Exercise 3

Solve an obtuse triangle having side a, side b, side c=6 cm, angle A=20 degrees, angle B and angle C=60 degrees.

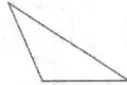

Solution:

Using the law of sines

First we will find the angle B of this obtuse triangle

Then **A + B + C=** 180 degrees

20 degrees + 60 degrees + B= 180 degrees

B= 180 degrees - 60 degrees - 20 degrees

B= 100 degrees

Angle B = 100 degrees

Then the a-side is found using the law of sines

$\dfrac{c}{sen\ c} = \dfrac{a}{sen\ a}$

$\dfrac{6}{sen\ 60\ grados} = \dfrac{a}{sen\ 20\ grados}$

Then **it remains** $\dfrac{6}{0.86} = \dfrac{a}{0.34}$

$\dfrac{6 \times 0.34}{0.86} = a$

$\dfrac{2.04}{0.86} = a$

2.37cm= a

Side b will be found using the law of sines

$\dfrac{c}{sen \ c} = \dfrac{b}{sen \ b}$

$\dfrac{6}{sen \ 60 \ grados} = \dfrac{b}{sen \ 100 \ grados}$

Then **it remains** $\dfrac{6}{0.86} = \dfrac{b}{0.98}$

$\dfrac{6 \times 0.98}{0.86} = b$

$\dfrac{5.88}{0.86} = b$

6.83cm= b

Answer: Side a is 2.37cm, side b is 6.83cm and angle b is 100 degrees.

Chapter 7: Solving an acute-angled triangle.

Definition of acute-angled triangle.

Acute-angled triangle

It is that triangle that has 3 angles that are acute, but are distinct.

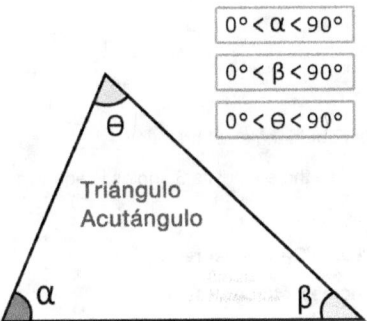

$$0° < α < 90°$$
$$0° < β < 90°$$
$$0° < θ < 90°$$

Triángulo
Acutángulo

7.2 Exercises solving an acute-angled triangle.

Exercise 1

Solve an acute-angled triangle having side a=10m, side b, side c, angle A= 60 degrees, angle B= 60 degrees and angle C.

Solution:

First the angle C of this right triangle will be found.

Knowing that in any triangle the sum of its 3 angles is equal to 180 degrees.

A +B + C= 180 degrees

60 degrees + 40 degrees + C= 180 degrees

C= 180 degrees - 60 degrees - 40 degrees

C= 80 degrees

Then, the side b will be found, using the law of sines

$$\frac{a}{sen\ a} = \frac{b}{sen\ b}$$

$$\frac{19}{sen\ 60\ grados} = \frac{b}{sen\ 40\ grados}$$

Then **it remains** $\frac{10}{0.86} = \frac{b}{0.64}$

$$\frac{10 \times 0.64}{0.86} = b$$

$$\frac{6.4}{0.86} = b$$

7.44m=b

The side c is found by reusing the law of sines

$$\frac{a}{sen\ a} = \frac{c}{sen\ c}$$

$$\frac{10}{sen\ 60\ grados} = \frac{c}{sen\ 80\ grados}$$

Then **it remains** $\frac{10}{0.86} = \frac{c}{0.98}$

$$\frac{10 \times 0.98}{0.86} = c$$

$$\frac{9.8}{0.86} = c$$

11.39m=c

Answer: Side b is 7.44m, side c is 11.39m and angle c is 80 degrees.

Exercise 2

Solve an acute-angled triangle having side a, side b, side c=5m, angle A=30 degrees, angle B and angle C=70 degrees.

Solution:

First the angle B of this right triangle will be found.

Knowing that in any triangle the sum of its 3 angles is equal to 180 degrees.

Then **A + B + C= 180 degrees.**

30 degrees + B + 70 degrees= 180 degrees

B= 180 degrees - 70 degrees - 30 degrees

B= 80 degrees

Next is the side a using sinus law

$$\frac{c}{sen\ C} = \frac{a}{sen\ A}$$

$$\frac{5}{sen\ 70\ grados} = \frac{a}{sen\ 30\ grados}$$

Then **it remains** $\dfrac{5}{0.93} = \dfrac{a}{0.5}$

$$\frac{5 \times 0.5}{0.93} = a$$

$$\frac{2.5}{0.93} = a$$

2.68cm= a

Side b is found, then it remains $\dfrac{c}{sen\ C} = \dfrac{b}{sen\ B}$

$$\frac{5}{sen\ 70\ grados} = \frac{b}{sen\ 80\ grados}$$

$$\frac{5}{0.93} = \frac{b}{0.98}$$

It remains that $\dfrac{5\ \times\ 0.98}{0.93} = b$

$$\frac{4.9}{0.93} = b$$

5.26cm= b

Answer: Side a is 5.68cm. side b is 5.26cm and angle b is 80 degrees.

Exercise 3

Solve an acute-angled triangle having side a, side b=12 cm, side c, angle A=60 degrees, angle B=50 degrees and angle C.

Solution:

First the angle C of the right triangle will be found as follows:

Knowing that in any triangle the sum of its 3 angles is equal to 180 degrees.

Then **A + B + C**= 180 degrees

60 degrees + 50 degrees + C=180 degrees.

C= 180 degrees - 60 degrees - 50 degrees

C=70 degrees

Find the side a of the right-angled triangle using the law of sines

$$\frac{b}{sen\ B} = \frac{a}{sen\ A}$$

$$\frac{12}{sen\ 50\ grados} = \frac{a}{sen\ 60\ grados}$$

Then it remains $\frac{12}{0.76} = \frac{a}{0.86}$

$$\frac{12\ \times\ 0.86}{0.76} = a$$

$$\frac{10.32}{0.76} = a$$

13.57= a

Then there is the side c

$$\frac{b}{sen\ B} = \frac{c}{sen\ C}$$

$$\frac{12}{sen\ 50\ grados} = \frac{c}{sen\ 70\ grados}$$

Then **it remains** $\frac{12}{0.76} = \frac{c}{0.93}$

$$\frac{12\ \times\ 0.93}{0.76} = c$$

$$\frac{11.16}{0.76} = c$$

14.68cm= c

Answer: Side a is 13.57cm. side c is 14.68cm and angle c is 70 degrees.

Chapter 8: The law of breasts

Definition of the law of sines.

Breast Law

It is the law that states that in any triangle the ratio of any of its sides to the sine of the opposite angle will always be constant. It should be noted that the law of sines is used most of the time in obtuse triangles and acute triangles.

Note that the law of sines can be used in any triangle, but it is used more in obtuse and acute triangles.

The law of sines is written as follows:

$$\frac{a}{sen\ A} = \frac{b}{sen\ B} = \frac{c}{Sen\ C}$$

These formulas work to find one side of the triangle.

Law of sines to find an angle in a triangle.

$$\frac{sen\ A}{a} = \frac{sen\ B}{b} = \frac{sen\ C}{c}$$

8.2 Exercises with the law of sines.

Example 1:

Find the side b, c and angle b of a triangle having side a=10 cm, angle A=30 degrees and angle c=70 degrees.

Solution:

First the missing angle B of this triangle will be found.

Knowing that in any triangle the sum of its 3 angles is equal to 180 degrees.

A + B + C= 180 degrees

30 degrees + B + 70 degrees= 180 degrees

B= 180 degrees - 70 degrees - 30 degrees

B= 80 degrees

Side c is found using the law of sines

$$\frac{a}{senA} = \frac{c}{senC}$$

$$\frac{10}{sen\ 30\ grados} = \frac{c}{sen\ 70\ grados}$$

Then **it remains** $\dfrac{10}{0.5} = \dfrac{c}{0.93}$

$$\frac{10 \times 0.93}{0.5} = c$$

$$\frac{9.3}{0.5} = c$$

18.6= c

Then we find the side b of this triangle

$$\frac{a}{sen\,A} = \frac{b}{sen\,B}$$

$$\frac{10}{sen\,30\ grados} = \frac{b}{sen\,80\ grados}$$

Then it remains $\dfrac{10}{0.5} = \dfrac{b}{0.98}$

$$\frac{10\ \times\ 0.98}{0.5} = b$$

Answer: Side b is 19.6cm, side c is 18.6cm and angle b is 80 degrees.

Example 2:

Find side a, side c and angle C of a triangle having side b=4 cm, angle A= 20 degrees and angle B= 100 degrees.

Solution:

First the angle B will be found

Knowing that in any triangle the sum of its 3 angles is equal to 180 degrees.

A + B + C= 180 degrees

20 degrees + 100 degrees + C= 180 degrees

C= 180 degrees - 100 degrees - 20 degrees

C= 60 degrees

We find the side a of this triangle using the law of sines

$$\frac{b}{senB} = \frac{a}{sen\,A}$$

$$\frac{4}{sen\,100\ grados} = \frac{a}{sen\ 20\ grados}$$

Then it remains $\dfrac{4}{0.98} = \dfrac{a}{0.34.}$

$$\frac{4\ \times\ 0.34}{0.98} = a$$

$$\frac{1.36}{0.98} = a$$

1.38cm= a

Then there is the side c

$$\frac{b}{sen\ B} = \frac{c}{senC}$$

$$\frac{4}{sen\ 100\ grados} = \frac{c}{sen\ 60\ grados}$$

Then **it remains** $\frac{4}{0.98} = \frac{c}{0.86}$

$$\frac{4 \times 0.86}{0.98} = c$$

$$\frac{3.44}{0.98} = c$$

3.51cm= c

Answer: Side a is 1.39cm, side c is 3.51cm and angle c is 60 degrees.

Example 3:

Find the value of x in a triangle having side a=14 cm, angle A=30 degrees and angle 100 degrees.

Solution:

The value of x is found as follows:

Using the law of sines.

$$\frac{14}{sen\ 30\ grados} = \frac{x}{sen\ 100\ grados}$$

$$\frac{14 \times 0.98}{0.5} = x$$

Then **it remains** $\frac{13.72}{0.5} = x$

27.44cm=x

The value of x in this triangle= 27.44cm

Example 4:

Find the angle C of a triangle having side a=40cm, angle A=42 degrees and side c=52cm.

Solution:

Angle C is found as follows:

Using the law of sines.

$$\frac{sen\ A}{a} = \frac{sen\ C}{c}$$

$$\frac{sen\ 42\ grados}{40} = \frac{sen\ C}{52}$$

$$\frac{52 \times sen\ 42\ grados}{40} = sen\ C$$

Then it is $\dfrac{52 \times sen^{-1}(42)\ grados}{40}$ = Sen⁻¹ C

60.44 degrees= C

Answer: Angle C is 60.44 degrees.

Example 5:

Find the angle C of a triangle having side a=16m, angle A=115 degrees and side c=10m.

Solution:

Angle C is found as follows:

Using the law of sines.

$$\frac{sen\ A}{a} = \frac{sen\ C}{c}$$

$$\frac{sen\ 115\ grados}{16} = \frac{sen\ C}{10}$$

$$\frac{10 \times sen\ 115\ grados}{16} = \sin C$$

Then **it remains** $\dfrac{10 \times sen^{-1}(115)\ grados}{16}$ = sen⁻¹ C

34.5 degrees= C

Answer: Angle C is 34.5 degrees.

 PA

Exercise 1

Find the side b of a triangle having angle A=50 degrees, angle B=30 degrees and side a=10cm.

Solution:

Side b is found as follows:

Using the law of sines

$$\frac{a}{sen\ A} = \frac{b}{sen\ B}$$

$$\frac{10}{sen\ 50\ grados} = \frac{b}{sen\ 30\ grados}$$

Then **it remains** $\dfrac{10}{0.76} = \dfrac{b}{0.5}$

$$\frac{10 \times 0.5}{0.76} = b$$

$$\frac{5}{0.76} = b$$

6.57cm= b

Answer: Side b is 6.57cm

Exercise 2

Find the side a of a triangle having side b=25 cm, angle A=42 degrees and angle B=10 degrees.

Solution:

How it is solved is as follows:

Using the law of sines.

$$\frac{b}{sen\ B} = \frac{a}{sen\ A}$$

$$\frac{25}{sen\ 10\ grados} = \frac{a}{sen\ 42\ grados}$$

Then **it remains** $\frac{25}{0.17} = \frac{b}{0.66}$

$$\frac{25\ \times\ 0.66}{0.17} = a$$

$$\frac{16.5}{0.17} = a$$

97.05cm= a

Answer: Side a is 97.05cm

Exercise 3

Find the side c of a triangle having side a=12 cm, angle A=20 degrees, angle C=30 degrees.

Solution:

Side c is found as follows:

Using the law of sines.

$$\frac{a}{sen\ A} = \frac{c}{sen\ C}$$

$$\frac{12}{sen\ 20\ grados} = \frac{c}{sen\ 30\ grados}$$

Then **it remains** $\dfrac{12}{0.34} = \dfrac{c}{0.5}$

$\dfrac{12 \ \times \ 0.5}{0.34} = c$

$\dfrac{6}{0.34} = c$

17.64cm= c

Answer: Side c is 17.64cm

Exercise 4

Find the side c of a triangle having side b=50 m, angle B=70 degrees, angle C=100 degrees.

Solution:

Side c is found as follows:

Using the law of sines

$\dfrac{b}{sen \ B} = \dfrac{c}{sen \ C}$

$\dfrac{50}{sen \ 70 \ grados} = \dfrac{c}{sen \ 100 \ grados}$

Then **it remains** $\dfrac{50}{0.93} = \dfrac{c}{0.98}$

$\dfrac{50 \ \times \ 0.98}{0.73} = c$

$\dfrac{49}{0.73} = c$

67.12m= c

Answer: Side c is 67.12m

Exercise 5

Find the side a of a triangle which has side b=4m, angle B=15 degrees, angle A=30 degrees.

Solution:

It is resolved as follows:

Using the law of sines.

$$\frac{b}{sen\ B} = \frac{a}{sen\ A}$$

$$\frac{4}{sen\ 15\ grados} = \frac{a}{sen\ 30\ grados}$$

Then **it remains** $\dfrac{4}{0.25} = \dfrac{a}{0.5}$

$$\frac{4\ \times\ 0.5}{0.25} = a$$

$$\frac{2}{0.25} = a$$

8m= a

Answer: Side a is 8m

8.3 Problems with the law of sines.

Problem 1:

A person is 6m from his house, on the roof of his house is a flag. The person observes the base of the flagpole at an elevation angle of 30 degrees and at the top of the flagpole is at an angle of 70 degrees. Determine the height of the flagpole.

Solution:

At the base of the flagpole is formed a right triangle having an angle of 30 degrees and one side of this right triangle having an adjacent cathetus measuring 6m.

Then to find the opposite leg of the right triangle we will use the trigonometric ratios.

This leaves **Cos=** $\dfrac{cateto \quad adyacente}{hipotenusa}$

Cos 30 degrees= $\dfrac{6}{c}$

C × cos 30 degrees= 6

C= $\dfrac{6}{0.86}$

C= 6.97m

Therefore, the hypotenuse of the right triangle formed is 6.97m.

- **Angle A=** 60 degrees
- **Angle B=** 30 degrees
- **Angle C=** 90 degrees

Then the other triangle will be used.

- **Angle A=** 40 degrees
- **Side c=** 6.97m
- **Angle B=** 120 degrees

The sine law will be used to determine the height of the flagpole.

Angle C will be found

A + B + C= 180 degrees

40 degrees + 120 degrees + C= 180 degrees

C= 180 degrees - 120 degrees - 40 degrees

C= 20 degrees.

$$\frac{c}{sen \ C} = \frac{a}{sen \ A}$$

$$\frac{6.97}{sen \ 20 \ grados} = \frac{a}{sen \ 40 \ grados}$$

PA

Then **it remains** $\dfrac{6.97}{0.34} = \dfrac{a}{0.64}$

$\dfrac{6.97 \times 0.64}{0.34} = a$

$\dfrac{44608}{0.34} = a$

13.12m= a

Answer: The height of the pole is 13.12m.

Chapter 9: The law of cosines.

Definition of cosine law.

Law of cosines

In a triangle the square of each side will be equal to the sum of the squares of the other two, minus twice the product of both by the cosine of the angle they form.

a^2= b² + c² - 2(b)(c) × Cos A

Formulas of the law of cosines.

This formula works for finding side a of an oblique triangle.

$$a^2= b² + c² - 2(b)(c) × Cos A$$

This formula works for finding side b of an oblique triangle.

$$b^2= a² + c² - 2(a)(c) × Cos B$$

This formula works to find the side c of an oblique triangle.

$$c^2= a² + b² - 2(a)(b) × Cos C$$

The law of cosines can be used in any triangle, but most of the time it is used in oblique triangles.

9.2 Exercises with the law of cosines.

Exercise 1

Find side b of a triangle having side a=12 cm, side c=7 cm, angle B =40 degrees.

Solution:

It is resolved as follows:

Using the law of cosines

b^2 = a² + c² - 2(a)(c) × cos B

b^2 = 12² + 7² - 2(12)(7) × cos 40 degrees.

$\sqrt{b^2} = \sqrt{(12^2 + 7^2 - 2 \times 12 \times 7 \times \cos 40\ grados)}$

B= 8.01cm

Remember that the sides of a triangle are named with lowercase letters.

Answer: Side b is 8.01cm

Exercise 2

Find the side a of a triangle having side b=13 cm, side c=9cm and angle A=120 degrees.

Solution:

It is resolved as follows:

Using the law of cosines.

a^2 = b² + c² - 2(b)(c) × cos A

a^2 = 13² + 9² - 2(13)(9) × cos 120 degrees.

$\sqrt{a^2} = \sqrt{(13^2 + 9^2 - 2 \times 13 \times 9 \times \cos 120\ grados)}$

A=19.45cm

Answer: Side a is 19.45cm

Exercise 3

Find the angle A of a triangle having side a=9cm, side b=7cm, side c=13cm.

Solution:

It is resolved as follows:

Using the law of cosines.

$$a^2 = b^2 + c^2 - 2(b)(c) \times \cos A$$

Then it is cleared and it is $a^2 - b^2 - c^2 = -2bc \cos A$

$$\frac{a^2 - b^2 - c^2}{-2bc} = \cos A$$

Cos $(^{-1} \dfrac{(9^2 - 7^2 - 13^2)}{-2 \times 7 \times 13}) =$ cos⁻¹ A

41.17 degrees= A

Answer: Angle A is 41.17 degrees.

Chapter 10: Operations with monomials.

Definition of monomial.

Monomial.

It is an algebraic expression in which the only operations that usually appear between the variables are the product and the power of the natural exponent.

Parts of a monomial.

- Coefficient
- Literal part
- Grade

Definition of the parts of a monomial

Coefficient.

The coefficient of a monomial is defined as the number that appears multiplying the literals.

Literal Part.

It is constituted by the letters and their exponents.

Grade.

The degree of a monomial is given by the sum of all the exponents of the variables.

Examples; find the degree of a monomial.

Example 1:

$2x^2$

R= The degree of this monomial is 2

Example 2:

$2x^2$ and ^{3}

R= The degree of this monomial is 5

Example 3:

$2b\ c^{25}$

R= The degree of this monomial is 7

Example 4:

$5x^3$ and x^{42}

R= The degree of this monomial is 10

Example 5:

$2w\ z^2$

R= The degree of this monomial is 3

Similar monomials

Two monomials are similar if they have the same literal part.

Examples of similar monomials

Example 1:

$2x^2$, $3x^2$ if it is an example of similar monomials.

Example 2:

$2z^3$ and $16z^3$ if it is an example of similar monomials.

Example 3:

$5c^2$ and $10c^2$ if it is an example of similar monomials.

Example 4:

$2x^4$, and^4 is not an example of similar monomials.

Example 5:

X^3 y $2z^5$ is not an example of similar monomials.

10.2 Addition of a monomial.

Sum of monomials

The sum of two monomials that have the same literal part and whose coefficients.
Only similar monomials can be added.

Examples of addition of a monomial

Example 1:

$5x + 2x = 7x$

Solution:

First the sum of monomials is observed if it has like terms and then the coefficients are added.

$5 + 2 = 7$

Answer: 7x

Example 2:

$2x^2 + 3x^2 = 5x^2$

Solution:

First we observe if the sum of monomials is similar, then we add the coefficients.

$2 + 3 = 5$

Answer: $5x^2$

Example 3:

$10x^2 x + 2x^2 5x = 12x 6x^2$

Solution:

First we look for whether the sum of monomials has like terms, then we add

$10x^2 + 2x^2 = 12x^2$

Then there is the other

$X + 5x = 6x$

Answer: $12x 6x^2$

Example 4:

$5b + d + 3b + 2d = 8b + 3d$

Solution:

First, we observe if there are like terms and if there are, we can perform the addition with monomials.

5b + 3b= 8b

D + 2d= 3d

8b + 3d

Answer: 8b +3d

Example 5:

2b + 3c + 4z + 5b + 2c= 7b + 5c + 4z

Solution:

First we observe if there are like terms to be able to add the monomials. In this case, if there are like terms and the ones that are not like terms remain the same, they cannot be added.

Then it is **2b + 5b= 7b**

3c + 2c= 5c

The 4z remains the same.

7b + 5c + 4z

Answer: 7b + 5c + 4z

10.3 Subtraction of a monomial.

Subtraction of monomials

It is to subtract an algebraic expression with only one term that is similar in order to solve it.

Examples of subtraction of a monomial

Example 1:

5x - 2x= 3x

Solution:

First we observe if there are like terms in the monomials. In this case as if like terms are found.

5 -2 = 3

Then add the x

Answer: 3x

Example 2:

10b - 4c - 6b - 5c= 4b - 9c

Solution:

First, the like terms are located so that they can be solved.

In this case there are like terms.

10b - 6b= 4b

-4c - 5c= -9c

4b - 9c

Answer: 4b - 9c

Example 3:

5b - 12c - 10a - 3b - 20a - 10c= -30a + 2b -22c

Solution:

First we observe if there are like terms in order to solve the subtraction.

Then it remains

-10a - 20a= -30a

5b - 3b= 2b

-12c - 10c= -22c

Then it remains

-30 a + 2b - 22c

Answer: -30a + 2b - 22c

10.4 Multiplication of a monomial.

Multiplication of monomials

It is another monomial whose coefficient is the product of the coefficients and whose literal part will be obtained by multiplying the powers that have the same base. The exponents would be added.

Examples; multiplication of monomials

Example 1:

$2X^2 \times x^8 = 2x^{10}$

Solution:

First, it is observed if equal bases are being multiplied.

$2 \times 1 = 2$

Then the exponents in the bases are added.

$2x^{10}$

Answer: $2x^{10}$

Example 2:

$5x \times 2x = 10x^2$

Solution:

First, it is observed whether there are equal bases in the monomials

$5 \times 2 = 10$

The exponents of the bases are then added together

$10x^2$

Answer: $10x^2$

Example 3:

$2x \ z^{32} \times 3x \ z^{43} = 6x \ z^{75}$

Solution:

First, it is observed if they have equal bases.

$2 \times 3 = 6$

Then the exponents of the bases are added together

$6x \ z^{75}$

Answer: $6x\ z^{75}$

Example 4:

$2c^2\ 3b\ 5\ a^3 \times a^5\ 2b^2\ 3c^6 = a^8\ 6b^3\ 6c^8$

Solution:

First, it is observed if there are equal bases.

$a^3 \times a^5 = a^8$

$3b \times 2b^2 = 6b^3$

Then it remains

$2c^2 \times 3c^6 = 6^8$

Then it is $a^8\ 6b^3\ 6c^8$

Answer: $a^8\ 6b^3\ 6c^8$

Example 5:

$3y^5 \times 5y^4 = 15y^9$

Solution:

First, it is observed whether they have the same bases.

$3 \times 5 = 15$

Then the exponents are added.

$15y^9$

Answer: $15y^9$

10.5 Division of a monomial.

Division of monomials

You can only divide monomials with the same literal part and with the degree of the dividend greater than or equal to the degree of the corresponding variable of the divisor.

Division of monomials

It is another monomial whose coefficient is the quotient of the coefficients whose literal part will be obtained by dividing the powers that have the same bases since it means that the exponents will be subtracted.

Examples of a division of monomials

Example 1:

$$\frac{x^{10}}{x^7} = \quad \textbf{x}^{\textbf{3}}$$

Solution:

First, it is observed if there are equal bases.

Then it remains

X³

Answer: X ³

Example 2:

$$\frac{20x^5}{5x^2} = \quad \textbf{4x}^{\textbf{3}}$$

Solution:

The first step is to check whether the bases are the same.

Then it remains

$$\frac{20}{5} = 4$$

The exponents are then subtracted

4x ³

Answer: 4x ³

Example 3:

$$\frac{10x^6}{5x^4} = 2x^2$$

Solution:

The first step is to check whether the same bases

Then it remains

$$\frac{10}{5} = 2$$

The exponents are then subtracted

2x ^{2}

Answer: 2x ^{2}

Example 4:

$$\frac{30b^5 \; c^8}{3b^2 \; c^3} = \quad \mathbf{10b^3 \; c^5}$$

Solution:

First, it is observed whether they have the same bases

Then it remains

$$\frac{30}{3} = 10$$

The exponents are then subtracted

10b^3 c ^{5}

Answer: 10b^3 c ^{5}

Example 5:

$$\frac{100x^{10} \; y^6 \; z^{12}}{20x^3 \; y^2 \; z^4} = \quad \mathbf{5x^7 \; y^4 \; z^8}$$

Solution:

First, it is observed if there are equal bases

Then it remains

$$\frac{100}{20} = 5$$

The exponents are then subtracted

$5x^7\ y^4\ z\ ^8$

Answer: $5x^7$ and^4 $z\ ^8$

10.6 Monomial Exercises

Exercise 1:

$8x + 2x = 10x$

Exercise 2:

$5x^2 + 3x^2 = 8x\ ^2$

Exercise 3:

$2x\ z^{32} + 3x^3\ z^2 = 5x\ z\ ^{32}$

Exercise 4:

$2b + 3c + 5^a + 3b + 4c + 2a = 7a + 5b + 7c$

Exercise 5:

$-2x^3 + 7x^3 = 5x\ ^3$

Exercise 6:

$12x - 10x = 2x$

Exercise 7:

$12x^2 - 8x^2 = 4x\ ^2$

Exercise 8:

$3x^3 \, y^2 \, z^4 - 10x^3 \, y^2 \, z^4 = -7x \, y \, z^{324}$

Exercise 9:

$-5x^4 - 3x^4 = -8x^4$

Exercise 10:

$2x \times 4x = 8x^2$

Exercise 11:

$5x^3 \times 8x^7 = 40x^{10}$

Exercise 12:

$5x^2 \times 3x^4 = 15x^6$

Exercise 13:

$6x \, z^{23} \times 4x^4 \, z^2 = 24x \, z^{65}$

Exercise 14:

$$\frac{20x^6}{2x^4} = 10x^2$$

Chapter 11: Operations with binomials

11.1 Definition of what a binomial is.

Binomial
It is a polynomial consisting of 2 monomials.

Examples of what a binomial is.

$P(x)= 2x^2 + 5$

$P(X)= 4X^3 + 6X^2$

$P(X)= 5X^3 + 10X^2$

11.2 Binomial exercises

Example 1:

$(x + y) \times (x + y) = x^2 + 2xy + y^2$

Solution:

First take the first term of the first parenthesis and multiply it by the entire term of the second parenthesis.

$X \times x = x^2$

Then it remains

$X \times y = xy$

The second term of the first parenthesis is taken and multiplied by all the terms of the second parenthesis.

Then it remains

$Y \times x \ xy = xy$

$Y \times y = y^2$

The like terms are then added if there are

$X^2 + xy + xy + y^2$

$X^2 + 2xy + y^2$

Answer: $X^2 + 2xy + y^2$

Example 2:

$(2x + 5z) \times (4x + 3z) = 8x^2 + 26xz + 15z^2$

Solution:

First, the first term of the first parenthesis is taken and multiplied by all the terms of the second parenthesis.

Then it remains

$2x \times 4x = 8x^2$

$2x \times 3z = 6xz$

The second term of the first parenthesis is taken and multiplied by all the terms of the second term.

$5z \times 4x = 20xz$

Then it remains

$5z \times 3z = 15z^2$

The like terms are added if there are any.

$8x^2 + 6xz + 20xz + 15z^2$

$8x^2 + 26xz + 15z^2$

Answer: $8x^2 + 26xz + 15z^2$

Example 3:

$(x + z) \times (2x + z) = 2x^2 + 3xz + z^2$

Solution:

First the first term of the first parenthesis is used and multiplied by all the terms of the second parenthesis.

Then it remains

$X \times 2x = 2x\ 2x^2$

$X \times z = xz$

The second term of the first parenthesis is taken and multiplied by all the terms of the second parenthesis.

Z × 2x= 2xz

Z × z= z ^{2}

Then the like terms are added if there are any.

Then it remains

$2x^2 + xz + 2xz + 2xz + z^2$

$2x^2 + 3xz + z^2$

Answer: $2x^2 + 3xz + z^2$

Example 4:

$(3x - x) × (2x - 5x) = -6x^2$

Solution:

First, the first term of the first parenthesis is taken and multiplied by all the terms of the second parenthesis.

Then it remains

$3x × 2x= 6x^2$

$3x × -5x= -15x^2$

The second term of the first parenthesis is taken and multiplied by all the terms of the second parenthesis.

$-x × 2x= -2x^2$

$-x × -5x= 5x^2$

The like terms are then subtracted or added if there are similar terms.

Then it remains

$6x^2 - 15x^2 - 2x^2 + 5x^2$

$6x^2 - 15x^2 = -9x^2$

Then it remains

$-9x^2 - 2x^2 = -11x^2$

$-11x^2 + 5x^2 = -6x^2$

Answer: -6x ^{2}

Example 5:

$(x - y) \times (x - y) = x^2 - 2xy + y^2$

Solution:

First, the first term of the first parenthesis is taken and multiplied by all the terms of the second parenthesis.

Then it remains

$X \times x = x^2$

$X \times -y = -xy$

The second term of the first parenthesis is taken and multiplied by all the terms of the second parenthesis.

Then it remains

$-y \times x = -xy$

$-y \times -y = y^2$

The like terms are then subtracted or added if there are similar terms.

$X^2 - xy - xy + y^2$

$X^2 - 2xy + y^2$

Answer: $X^2 - 2xy + y^2$

Exercise 1:

$(x + 5) \times (x + 5) = x^2 + 10x + 25$

Exercise 2:

$(2x + 2) \times (x + 2) = 2x^2 + 6x + 4$

Exercise 3:

$(2x + y) \times (5x + y) = 10x^2 + 7xy + y^2$

Exercise 4:

$(4x + 1) \times (3x + 1) = 12x^2 + 7x + 1$

Exercise 5:

$(xy + 3) \times (xy + 3) = x\,y^{22} + 6xy + 9$

Exercise 6:

$(x - 2) \times (x - 2) = x^2 - 4x + 4$

Exercise 7:

$(2x - 1) \times (5x - 1) = 10x^2 - 7x + 1$

Exercise 8:

$(x - w) \times (x - w) = x^2 - 2xw + w^2$

Exercise 9:

$(y - 4) \times (y - 4) = y^2 - 8y + 16$

Chapter 12: Operations with trinomials

12.1 Definition of what is a trinomial.

Trinomial

It is a polynomial that has 3 monomials.

Example of a trinomial

$P(x) = 3x^2 + 2x + 5$

Types of trinomials

- Perfect square trinomial

- Trinomial **squared**

- Second degree trinomial

Definition of trinomial types

Perfect square trinomial

It is the development of a binomial squared.

$(a + b)^2 = a^2 + 2ab + b^2$

Example:

$X^2 + 10x + 25 = (x + 5)^2$

Trinomial squared

It is equal to the square of the first plus the square of the second plus the square of the third + the double of the first times the second + the double of the first times the third plus the double of the second times the third.

$(a + b + c)^2 = a^2 + b^2 + c^2 + 2 \times a \times b + 2 \times a \times c + 2 \times b \times c$

Example:

$(3x + y + 4)^2$

Solution:

$3x^2 + y^2 + 4^2 + 2(3x)(y) + 2(y)(4) + 2(3x)(4)$

Then it remains

$9x^2 + y^2 + 16 + 6xy + 8y + 24x$

$9x^2 + y^2 + 6xy + 8y + 24x + 16$

Answer: $9x^2 + y^2 + 6xy + 8y + 24x + 16$

Second degree trinomial

In order to decompose a trinomial of second degree

$P(x) = ax^2 + bx + c$

Equal to 0 and solve the second degree equation.

Example:

$X^2 + 10x + 9$

This second degree trinomial equals 0

Then it remains

$X^2 + 10x + 9 = 0$

This second degree equation will be solved by the general formula method.

The equation is as follows:

$Ax^2 + bx + c = 0$

$A = 1, B = 10, C = 9$

The general formula is then used

$$X = \frac{-b \pm \sqrt{b^2 - 4(a)(c)}}{2a}$$

$$X = \frac{-10 \pm \sqrt{10^2 - 4(1)(9)}}{2}$$

$$X = \frac{-10 \pm \sqrt{100 - 36}}{2}$$

This leaves $X = \dfrac{-10 \pm \sqrt{64}}{2}$

$$X= \frac{-10 \ + \ 8}{2}$$

$$X= \frac{-10 \ - \ 8}{2}$$

$$X= \frac{-2}{2}$$

$$X= \frac{-18}{2}$$

Then it remains

X1= -1

X2= -9

Then it remains

(x - x1) × (x - x2)

(x - 1) × (x - 9)

Answer: (x - 1) × (x - 9)

12.2 Trinomial exercises.

Exercises perfect square trinomial.
Exercise 1
$x^2 + 6x + 9$
Solution:
First take the square root of x^2 and then take the square root of 9.
Then open a square parenthesis.

$$\sqrt{x^2} \ = \ x$$

This leaves $\sqrt{9} \ = \ 3$
$(x + 3)^2$
 Answer: (x + 3) ^{2}

Verification.

If solving the binomial squared gave you the answer a perfect squared trinomial, the answer will be correct.
Then it is
(x + 3) × (x + 3)

x × x= x ^{2}
Then it is
x × 3= 3x
3 × x= 3x
3 × 3 = 9
Then it is left
x^2 + 3x + 3x + 3x + 9
x^2 + 6x + 9
The final conclusion is that the answer is correct.

Exercise 2
x^2 + 4x + 4
Solution:
First take the square root of x^2 and then take the square root of 4.
Then open a square parenthesis.
Then

you have $\sqrt{x^2} = x$ $\sqrt{4} = 2$
(x + 2)^2
 Answer: (x + 2) ^{2}

Check
If when solving the binomial squared the answer gives the perfect squared
trinomial it means that it is correct.
Then it is
(x + 2) × (x + 2)
x × x= x ^{2}
x × 2= 2x
Then it is
2 × x= 2x
2 × 2= 4
x^2 + 2x + 2x + 4
x^2 + 4x + 4
The final conclusion is that the answer is correct.

Exercise 3
x^2 + 20x + 100
Solution:
First square root x^2 and then square root 100, then square open parentheses.

Then it remains

$$\sqrt{x^2} = x$$
$$\sqrt{100} = 10$$

Then it is

$(x + 10)^2$

Answer: $(X + 10)^2$

Check

If when solving the binomial squared the answer gives the perfect squared trinomial then the answer is correct.

Then it is

$(x + 10) \times (x + 10)$

$x \times x = x^2$

Then

$x \times 10 = 10x$

$10 \times x = 10x$

$10 \times 10 = 100$

$x^2 + 10x + 10x + 100$

$x^2 + 20x + 100$

The final conclusion is that the answer is correct.

Exercise 4

$x^2 + 14x + 49$

Solution:

First take the square root of x^2 and 49, then open the parentheses to the square.

Then

it is $\sqrt{x^2} = x$

$\sqrt{49} = 7$ That leaves $(x + 7)^2$

Answer: $(x + 7)^2$

Check

If when solving the binomial squared the answer gives the perfect squared trinomial it means that the answer is correct.

Then

$(x + 7) \times (x + 7)$

$x + x = x^2$

$x \times 7 = 7x$

$7 \times x = 7x$

$7 \times 7 = 49$

Then there is

$X^2 + 7x + 7x + 49$

$x^2 + 14x + 49$

The final conclusion is that the answer is correct.

Exercises; trinomial squared.

Exercise 1

$(x + y + z)^2$

Solution:

Using the a of a trinomial squared.

$a^2 + b^2 + c^2 + 2(a)(b) + 2(a)(c) + 2(b)(c)$

Then it remains

$X^2 + y^2 + z^2 + 2(x)(y) + 2(x)(z) + 2(y)(z)$

$X^2 + y^2 + z^2 + 2xy + 2xz + 2yz$

Answer: $X^2 + y^2 + z^2 + 2xy + 2xz + 2yz$

Exercise 2

$(2x + 3y + 5)^2$

Solution:

Using the formula of a trinomial squared.

$a^2 + b^2 + c^2 + 2(a)(b) + 2(a)(c) + 2(b)(c)$

$2x^2 + 3y^2 + 5^2 + 2(2x)(3y) + 2(2x)(5) + 2(3y)(5)$

Then it remains

$4x^2 + 9y^2 + 25 + 12xy + 20x + 30y$

$4x^2 + 9y^2 + 12xy + 30y + 20x + 25$

Answer: $4x^2 + 9y^2 + 12xy + 30y + 20x + 25$

Exercise 3

$(x + 2x + 1)^2$

Solution:

Using the trinomial squared formula.

$a^2 + b^2 + c^2 + 2(a)(b) + 2(a)(c) + 2(b)(c)$

Then it remains

$X^2 + 2x^2 + 1^2 + 2(x)(2x) + 2(x)(1) + 2(2x)(1)$

$X^2 + 4x^2 + 1 + 4x^2 + 2x + 4x$

$X^2 + 4x^2 + 4x^2 = 9x^2$

$2x + 4x = 6x$

$9x^2 + 6x + 1$

Answer: $9x^2 + 6x + 1$

Exercise 4

$(x + x + 5)^2$

Solution:

$X^2 + x^2 + 5^2 + 2(x)(x) + 2(x)(5) + 2(x)(5)$

Then it remains

$X^2 + x^2 + 25 + 2x + 2x^2 + 10x + 10x + 10x$

$X^2 + x^2 + 2x^2 = 4x^2$

$10x + 10x = 20x$

$4x^2 + 20x + 25$

Answer: $4x^2 + 20x + 25$

Exercise 5

$(2y + y + 3)^2$

 Solution:

$2y^2 + y + 3^{22} + 2(2y)(y) + 2(2y)(3) + 2(y)(3)$

Then there is

$4y^2 + y^2 + 9 + 4y^2 + 12y + 6y$

$4y^2 + y^2 + 4y^2 = 9y^2$

 $12y + 6y = 18y$

$9y^2 + 18y + 9$

Answer: 9y² + 18y + 9

Exercises of trinomial of second degree.

Exercise 1

X² + 2x + 1
Solution:

First the second degree trinomial is equaled to 0 and then solved as a second degree equation.

Then it remains

X² + 2x + 1= 0

This second degree equation is of the following form:

Ax² + bx + c= 0

A= 1, B=2, C=1

The general formula is used.

$$x = \frac{-b \pm \sqrt{b^2 - 4(a)(c)}}{2a}$$

$$X = \frac{-2 \pm \sqrt{2^2 - 4(1)(1)}}{2}$$

$$X = \frac{-2 \pm \sqrt{4 - 4}}{2}$$

$$X = \frac{-2 \pm \sqrt{0}}{2}$$

$$X = \frac{-2 \pm 0}{2}$$

$$X = \frac{-2 + 0}{2}$$

$$X = \frac{-2 - 0}{2}$$

$$X= \frac{-2}{2}$$

$$X= \frac{-2}{2}$$

Then it is

x1= -1

x2= -1

(x - 1) × (x - 1)

Answer: (x - 1) × (x - 1)

Exercise 2

X² + 6x + 5

Solution:

First the second degree trinomial is equated to 0 and then the equation is solved.

This leaves

X² + 6x + 5= 0

This second degree equation is in the following form:

Ax² + bx + c= 0

A= 1, B=6, C=5.

The general formula will be used.

$$X= \frac{-b \pm \sqrt{b^2 - 4(a)(c)}}{2a}$$

$$X= \frac{-6 \pm \sqrt{6^2 - 4(1)(5)}}{2}$$

$$X= \frac{-6 \pm \sqrt{36 - 20}}{2}$$

$$X= \frac{-6 \pm \sqrt{16}}{2}$$

$$X= \frac{-6 \pm 4}{2}$$

$$x= \frac{-6 + 4}{2}$$

$$x= \frac{-6 - 4}{2}$$

$$x= \frac{-6 - 4}{2}$$

$$X= \frac{-2}{2}$$

$$X= \frac{-10}{2}$$

Then it is

x1= -1
x2= -5
(x - 1) × (x - 5)
Answer: (x - 1) × (x - 5)

Exercise 3^{X2} **+ 15x + 50**
Solution:
First the second degree trinomial is equated to 0 and then the equation is solved.
Then it is
X^2 + 15x + 50= 0
This second degree equation is in the following form:
Ax^2 + bx + c= 0
A=1, B=15, C=50.
The general formula is used.

$$X= \frac{-b \pm \sqrt{b^2 - 4(a)(c)}}{2a}$$

$$X= \frac{-15 \pm \sqrt{15^2 - 4(1)(50)}}{2}$$

$$X= \frac{-15 \pm \sqrt{225 - 200}}{2}$$

$$X= \frac{-15 \pm \sqrt{25}}{2}$$

$$X= \frac{-15 \pm 5}{2}$$

$$X= \frac{-15 + 5}{2}$$

$$X= \frac{-15 - 5}{2}$$

$$X= \frac{-10}{2}$$

$$X= \frac{-20}{2}$$ Then

x1= -5
x2= -10

(x - 5) × (x - 10)
Answer: (x - 5) × (x - 10)

Chapter 13: Operations with polynomials.

13.1 Definition of what a polynomial is.

Polynomial

An algebraic expression that is made up of one or more terms using only the arithmetic operations of addition, subtraction, multiplication, division and exponents.

An algebraic expression that has only one term is called a monomial.

An algebraic expression with two terms is called a binomial.

The algebraic expression that has three terms is called a trinomial.

Addition of polynomials

The sum of polynomials refers to the combination of terms similar to those with the same degree of exponent.

Examples of polynomial sums

Example 1:
(2x + 1) + (4x + 2)

First the parentheses are eliminated and then the like terms are placed.

Then it remains

2x + 1 + 4x + 2

Then the like terms are added together.

2x + 4x= 6x

1+2= 3

6x + 3

Answer: 6x + 3

Example 2:

(4x + 5) + (6x + 3)

First the parentheses are removed and then the like terms are grouped together.

Then it remains

4x + 5 + 6x + 3

4x + 6x= 10x

Then it remains

5 +3=8

10x + 8

Answer: 10x + 8

Example 3:

(5x + 2y) + (7x + 3y)

First the parentheses are eliminated and then the like terms are grouped together.

Then it remains

5x + 2y + 7x + 3y

5x + 7x= 12x

Then it remains

2y + 3y= 5y

12x + 5y

Answer: 12x + 5y

Example 4:

(8x + 5y) (10x + 2y)

First the parentheses are eliminated and then the like terms are grouped together.

Then it remains

8x + 5y + 10x + 2y

8x + 10x= 18x

Then it remains

5y + 2y= 7y

18x + 7y

Answer: 18x + 7y

Example 5:

$(x^2 + y^2 + z) + (2x^2 + 5y)^2$

First the parentheses are eliminated and then the like terms are grouped together.

Then it remains

$X^2 + y^2 + z + 2x^2 + 5y^2$

$X^2 + 2x^2 = 3x^2$

$Y^2 + 5y^2 = 6y^2$

Then it remains

z

$3x^2 + 6y^2 + z$

Answer: $3x^2 + 6y^2 + z$

Polynomial subtraction

It consists of adding the minuend to the opposite of the subtrahend.

Examples of polynomial subtraction

Example 1:

(3x + 1) - (5x - 3)

Solution:

The terms of the first polynomial remain the same and the terms of the second polynomial have their sign changed.

Then it remains

3x + 1 - 5x +3

3x - 5x= -2x

Then it remains

1+ 3= 4

-2x + 4

Answer: -2x + 4

Example 2:

(8x + 5) - (2x - 4)

Solution:

The terms of the first polynomial remain the same and the terms of the second polynomial have their sign changed.

Then it remains

8x + 5 - 2x + 4

8x - 2x= 6x

Then it remains

5 + 4= 9

6x + 9

Answer: 6x + 9

Example 3:

(4x + 5y) - (3x + 2y)

Solution:

The terms of the first polynomial remain the same and the terms of the second polynomial have their sign changed.

Then it remains

4x + 5y - 3x - 2y

4x - 3x= x

Then it remains

5y - 2y= 3y

X + 3y

Answer: X + 3y

Example 4:

(10x + 4y) - (-6x - 6y)

Solution:

The terms of the first polynomial remain the same and the terms of the second polynomial have their sign changed.

Then it remains

10x + 4y + 6x + 6y

10x + 6x= 16x

Then it remains

4y + 6y= 10y

16x + 10y

Answer: 16x + 10y

Example 5:

$(x^2 + y^2) - (5x^2 + 9y)^2$

Solution:

The terms of the first polynomial remain the same and the terms of the second polynomial have their sign changed.

Then it remains

$X^2 + y^2 - 5x^2 - 9y^2$

$X^2 - 5x^2 = -4x^2$

Then it remains

$Y^2 - 9y^2 = -8y^2$
$-4x^2 - 8y^2$

Answer: $-4x^2 - 8y^2$

Multiplication of a monomial by a polynomial

The monomial is multiplied by each and every monomial that can form the polynomial.

Examples of multiplication of a monomial by a polynomial.
Example 1:

$5 \times (2x + 3)$

Solution:
The monomial is multiplied by each term of the polynomial.

That leaves
$5 \times 2x = 10x$
$5 \times 3 = 15$
$10x + 15$

Answer: 10x + 15

Example 2:

2 × (10x + 4)

Solution:

The monomial is multiplied by each term of the polynomial.

Then it remains

2 × 10x= 20x

2 × 4=8

20x + 8

Answer: 20x + 8

Example 3:

X × (2x + 5y + 3)

Solution:

The monomial is multiplied by all the terms of the polynomial.

Then
$x \times 2x = 2x = 2x^2$

X × 5y= 5xy
x × 3= 3x

$2x^2 + 5xy + 3x$

Answer: $2x^2 + 5xy + 3x$

Example 4:

2x × (5x + 6)

Solution:
The monomial is multiplied by all the terms of the polynomial.
 Then it is

$2x \times 5x = 10x^2$
 2x × 6= 12x
$10x^2 + 12x$
Answer: $10x^2 + 12x$

Example 5:

$x^2 \times (x + y + z)$

Solution:

The monomial is multiplied by all the terms of the polynomial.

Then it remains

$X^2 \times x = x^2$

$X^2 \times y = x\,y^2$

Then it is

$x^2 \times z = x^2$

$z\,x^3 + x^2\,y + x\,z^2$

Answer: $x^3 + x^2\,y + x\,z^2$

Multiplication of polynomials

Multiply each monomial of the first polynomial by all the elements of the second polynomial, add the monomials of the same degree and obtain another polynomial whose degree is the sum of the degrees of the polynomials that were multiplied.

Examples of polynomial multiplication

Example 1:

$(x + y) \times (2x + 3y + 2)$

Solution:

First, the first term of the first polynomial is multiplied by all the terms of the second polynomial.

Then $X \times 2x = 2x = 2x^2$

$X \times 3y = 3xy$

$X \times 2 = 2x$

The second term of the first polynomial is multiplied by all the terms of the second polynomial.

$Y \times 2x = 2xy$

$Y \times 3y = 3y^2$

$Y \times 2 = 2y$

Then it remains

$2x^2 + 3xy + 2x + 2xy + 2xy + 3y^2 + 2y$

$2x^2 + 3y^2 + 3xy + 2xy + 2xy + 2y + 2x$

$2x^2 + 3y^2 + 5xy + 2y + 2x$

Answer: $2x^2 + 3y^2 + 5xy + 2y + 2x$

Example 2:

$(b + c) \times (3b + 4c)$

Solution:

The first term of the first polynomial will multiply all the terms of the second polynomial.

Then it remains

B× 3b= $3b^2$

B × 4c= 4bc

The second term of the first polynomial is multiplied by all the terms of the second polynomial.

Then it remains

C × 3b= 3bc

C × 4c= $4c^2$

$B^2 + 4bc + 3bc + 4c^2$

$B^2 + 7bc + 4c^2$

Answer: $B^2 + 7bc + 4c^2$

Example 3:

$(2x + 5y) \times (x + y)$

Solution:

The first term of the first polynomial will multiply all the terms of the second polynomial.

Then it remains

2x × x= $2x^2$

$2x \times y = 2xy$

The second term of the first polynomial is multiplied by all the terms of the second polynomial.

PA

Then it remains

$5y \times x = 5xy$

$5y \times y = 5y^2$

$2x^2 + 2xy + 2xy + 5xy + 5y^2$

$2x^2 + 7xy + 5y^2$

Answer: $2x^2 + 7xy + 5y^2$

Example 4:

$(x^2 + z^2) \times (3x + 10z)$

Solution:

First, the first term of the first polynomial is multiplied by all the terms of the second polynomial.

Then it remains

$X^2 \times 3x = 3x^3$

$X^2 \times 10z = 10x z^2$

The second term of the first polynomial is multiplied by all the terms of the second polynomial.

Then it remains

$Z^2 \times 3x = 3xz^2$

$Z^2 \times 10z = 10z^3$

$3x^3 + 10x^2 z + 3xz^2 + 10z^3$

Answer: $3x^3 + 10x^2 z + 3xz^2 + 10z^3$

Example 5:

$(2x + 3y) \times (5x^2 + 2y)$

Solution:

First, the first term of the first polynomial is multiplied by all the terms of the second polynomial.

Then it remains

2x × 5x² = 10x ³

2x × 3y= 6xy

The second term of the first polynomial is multiplied by all the terms of the second polynomial.

Then it remains

3y × 5x² = 15x y²

3y × 2y= 6y ²

10x³ + 6xy +15x² y + 6y ²

10x³ + 6y² + 15x² y + 6xy

Answer: 10x³ + 6y² + 15x² y + 6xy

Division of polynomial by a monomial
In an exact division of polynomials the remainder will be 0 and the integer division of polynomials its remainder will be different from 0.

Fundamental property of division says the following:
The degree of the residue polynomial is always less than the degree of the dividing polynomial.

Division of a polynomial by a monomial
Each monomial of the polynomial is divided until the degree of the dividend is less than the degree of the divisor.

Example 1: $\dfrac{x^5 + x^2y^2}{x^2}$

Solution

$\dfrac{x^5}{x^2} = $ x ³

Then

$$\frac{x^2 y^2}{x^2} =$$

it remains $\quad$ y

Remains **x³ + y²**

Answer: x³ + y ²

Example 2: $\quad \dfrac{x^6 + x^{10}}{x^4}$

Solution: $\dfrac{x^6}{x^4} =$ **x ²**

Then

it is $\quad \dfrac{x^{10}}{x^4} =$ **x ⁶**

x² + x⁶

Answer: x² + x ⁶

Example 3

: $\quad \dfrac{4x + 6x}{2x}$

Solution

: $\dfrac{4x}{2x} = 2$

$\dfrac{6x}{2x} = 3$

Answer: 2 +3

Example 4

: $\quad \dfrac{20y + 50y + 15y}{5y}$

Solution

: $\dfrac{20y}{5y} = 4$

$\dfrac{50y}{5y} =$

10 $\dfrac{15y}{5y} = 3$

4 + 10 + 3
Answer: 4 + 10 + 3

Example 5:

$$\frac{20x^8 \;+\; 10x^9}{2x^3}$$

Solution: $\frac{20x^8}{2x^3} =$ **10x** ^{5}

Then it remains

$\frac{10x^9}{2x^3} =$ **5x⁶**

10x⁵ + 5x⁶

Answer: 10x⁵ + 5x ^{6}

Examples of division of a polynomial by another polynomial.

Example1: $\frac{x^2 \;+\; x \;-\; 20}{x \;+\; 5}$

Solution

$\therefore$ $\frac{x^2}{x} = x$

The following is done:

x × -x = -x ^{2}

Then

x × -5= -5x

x² - x² = 0

X - 5x= -4x

0x² - 4x - 20

That leaves $\frac{-4x}{x} = -4$

-4 × -x= 4x

-4 × -5= 20

-4x + 4x= 0x

-20 + 20= 0

That leaves **x- 4**

Answer: x - 4

Example 2: $\frac{x^2 \;+\; x \;-\; 10}{x \;+\; 2}$

Solution: $\dfrac{x^2}{x}$ = x

Then we do the following:

x × -x= -x^2

x × -2 = -2x

x^2 - x^2 = 0x^2

x - 2x= -x

0x^2 - x - 10

$$\dfrac{-x}{x} = -1$$

Then

-1 × -x= x

-1 × -10= 10

-x + x= 0x

-10 + 10= 0

x - 1

Answer: x- 1

13.2 Polynomial exercises.

Exercise 1

(x + 2) + (3x + 5) = 4x+7

Exercise 2

(x^2 + x + y) + (5x + 6x^2) = 7x^2 + 6x + y

Exercise 3
$(10x^3 + 5) + (3 + 5x^3) = 15x^3 + 8$

Exercise 4
$(2x + 5) - (5x + 2) = -3x + 3$

Exercise 5
$(x + y) - (-6x + 2y) = 7x - y$

Exercise 6
$2x \times (x^2 + 5) = 2x^3 + 10x$

Exercise 7
$X^2 \times (5x^6\ 4 + 3x) = 5x^8 + 3x^3$

Exercise 8
$(x + y) \times (2x + 4) = 2x^2 + 4x + 2xy + 4y$

Exercise 9
$$\frac{30x^7 + 9x^{12} - 6x^5}{3x^2} = $$
$10x^5 + 3x^{10} - 2x^3 - 2x$

Chapter 14: Types of factorization

Difference of squares

Examples of the difference of squares
Example 1:
$x^2 - 25$
Solution:

First find the square root of x^2 y of 25, then open two parentheses, one with positive and one with negative.

$\sqrt{x^2} = x$

$\sqrt{25} = 5$

Then there is **(x + 5) (x -5)**

Answer: (x + 5) (x - 5)

Check

If you want to be really sure that your answer is correct then you will have to do the following:

(x + 5) (x - 5)

If solving this gives you the following

x^2 - 25

Then the conclusion you will come to is that your answer is correct.

(x + 5) (x - 5)
x × x= x^2
 x × -5= -5x
5 × x= 5x

Then it is

5 × -5= -25
x^2 - 5x + 5x + 25
x^2 - 25

So the answer is correct.

Example 2:
x^2 - 100
Solution:
First find the square root of x^2 and of 100, then open two parentheses one with

positive and the other with negative. $\sqrt{x^2} = x$

$\sqrt{100} = 10$

Then you are left with **(x + 10) (x - 10)**

Answer: (x + 10) (x - 10)

Checking
(x + 10) × (x - 10)
If solving this gives the following
x^2 -100
The answer will be correct.

This leaves (
x + 10) × (x - 10)

x × x= x²
x × -10= -10x
10 × x= 10x
10 × -10= -100
x² - 10x + 10x - 100
This leaves x² - 100
So the answer is correct.

Example 3:
x² - 9

Solution:
First find the square root of x² y of 9, then open two parentheses one with positive and one with negative.
Then
it is $\sqrt{x^2} = x$

$\sqrt{9} = 3$
(x + 3) (x - 3)
Answer: (x + 3) (x - 3)
Check:
(x + 3) × (x - 3)
If solving this gives the following
x² - 9
The answer will be correct.
 Then it is
(x + 3) × (x - 3)
x × x= x²
x × -3= -3x
3 × x= 3x
3 × -3= -9
x² - 3x + 3x - 9
This leaves x² - 9
The answer is correct.

Example 4:
x² - 4
Solution:
First find the square root of x² y of 4, then open two parentheses one with positive and one with negative.
Then
it is $\sqrt{x^2} = x$
$\sqrt{4} = 2$

(x + 2) (x - 2)
Answer: (x +2) (x- 2)
Check:
(x + 2) × (x - 2)
If solving this gives the following.
x² - 4
The answer will be correct.
Then it is:
(x + 2) × (x - 2)
x × x= x²
 x × -2= -2x
2 × x= 2x
2 × -2= -4
x² - 2x + 2x - 2x - 4
Then there remains **x² - 4**
So the answer is correct.

Example 5
x² - 36
Solution:
First find the square root of x² and of 36, then open two parentheses one with positive and the other with negative.
Then

it is $\sqrt{x^2} = x \sqrt{36} = 6$

 (x + 6) (x - 6)
Answer: (x + 6) (x - 6)
Check:
(x + 6) (x - 6)
If solving this gives the following
x² - 36
The answer will be correct.
Then
(x + 6) × (x - 6)
x × x= x²
 x × -6= -6x
6 × x= 6x
6 × -6= -36
x² - 6x + 6x - 36
So that leaves **x² - 36**
So the answer is correct.

Examples of difference of cubes.
Example 1:
$x^3 - 27$
Solution:
First find the cube root of x^3 and of 27, then open two parentheses.

This leaves $\sqrt[3]{x^3} = x$
$\sqrt[3]{27} = 3$
Then it is **(x - 3) (x² + 3x + 9)**
Answer: (x - 3) (x² + 3x+ 9)

Example 2:

X³ - 8
Solution:
First find the cube root of x^3 y of 8, then open two parentheses.
Then
it is $\sqrt[3]{x^3} = x$
$\sqrt[3]{8} = 2$
This leaves **(x - 2) (x² + 2x + 4)** **Answer: (x- 2) (x² + 2x + 4)**

Example 3:
x³ - 1000

Solution:

First find the cube root of x^3 y of 1000, then open two parentheses. $\sqrt[3]{x^3} = x$
$\sqrt[3]{1000} = 10$

Then it is **(x - 10) (x² + 10x + 100)**

Answer: (x - 10) (x² + 10x + 100)

Example 4:

X³ - 64

Solution:

First find the cube root of x^3 y of 64, then open two parentheses.

Then it remains

$\sqrt[3]{x^3} = x$

$\sqrt[3]{64} = 4$

This leaves **(x - 4) (x² + 4x + 16)**

Answer: (x - 4) (x² + 4x + 16)

Example 5:

X³ - 125

Solution:

First find the cube root of x^3 y of 125, then open two parentheses.

Then it remains.

$\sqrt[3]{x^3} = x$

$\sqrt[3]{125} = 5$

Then there is **(x - 5) (x² + 5x + 25)**

Answer: (x - 5) (x² + 5x + 25)

14.3 Sum of cubes

Examples of sum of cubes

Example 1:

X³ + 8

Solution:

First find the cube root of x^3 y of 8, then open two parentheses.

Then it remains

$\sqrt[3]{x^3} = x$

$\sqrt[3]{8} = 2$

(x + 2) (x² - 2x + 4)

Answer: (x + 2) (x² - 2x + 4)

 PA

Example 2:

X³ + 125

Solution:

First find the cube root of x³ y of 125, then open two parentheses.

Then it remains

$\sqrt[3]{x^3} = x$

$\sqrt[3]{125} = 5$

(x + 5) (x² - 5x + 25)

Answer: (x + 5) (x² - 5x + 25)

Example 3:

X³ + 1000

Solution:

First find the cube root of x³ y of 1000, then open two parentheses.

Then it remains

$\sqrt[3]{x^3} = x$

$\sqrt{1000} = 10$

(x + 10) (x² - 10x + 100)

Answer: (x + 10) (x² - 10x + 100)

Example 4:

X³ + 27

Solution:

First find the cube root of x³ y of 27, then open two parentheses.

Then it remains

$$\sqrt[3]{x^3} = x$$

$$\sqrt[3]{27} = 3$$

(x + 3) (x² - 3x + 9)

Answer: (x + 3) (x² - 3x + 9)

Example 5:

X³ + 64
Solution:

First find the cube root of x³ y of 64, then open two parentheses.

$$\sqrt[3]{x^3} = x$$

$$\sqrt[3]{64} = 4$$

(x + 4) (x² - 4x + 16)
Answer: (x + 4) (x² - 4x + 16)

14.4 Common factor

Examples of common factor

Example 1:
2x + 2y

Solution:

2(x + y)

Answer: 2(x + y)

Example 2:

10b - 15c

Solution:

Using the DCM, then the DCM of (10,15) will be found as follows:

The fifth will be taken from the 10 and 15

Then it remains

$$\frac{10}{5} = 2$$

$$\frac{15}{5} = 3$$

It then follows that the DCM of 10 and 15 is equal to 5.

5(2b - 3c)

Answer: 5(2b - 3c)

Example 3:

4pq + 8pr - 12pt

Solution:

First P will be taken out since it is the most repeated letter, then the DCM will be found.

The DCM of (4.8.12)

It is found as follows:

First half is taken from 4, then from 8 and finally from 12.

Then it remains

$$\frac{4}{2} =$$

$$2\frac{8}{2} = 4$$

$$\frac{12}{2} = 6$$

The following is done:
We remove half from 2, 4 and 6.
Then it is

$$\frac{2}{2} = 1$$

$$\frac{4}{2} = 2$$

$$\frac{6}{2} = 3$$

2 × 2= 4

So the DCM of (4,8,12) = **4**
4p (q + 2r +3t)

Answer: 4p (q + 2r +3t)

Example 4:
5x + 5z

Solution:
5 (x + z)
Answer: 5(x + z)

Example 5:
3x + 9y
Solution:
The DCM of 3 and 9 will be found first, then the third part will be taken from 3 and 9.
Then it is

$$\frac{3}{3} = 1$$
$$\frac{9}{3} = 3 \quad \text{DCM= 3}$$

Answer: 3(x + 3y)

14.5 Perfect square trinomial

Examples of perfect square trinomial

Example 1:
x^2 + 8x + 16
Solution:
First find the square root of x^2 and of 16, then open a square bracket.

Then

it remains $\sqrt{x^2} = x$
$\sqrt{16} = 4$
$(x + 4)^2$
 Answer: $(x + 4)^2$

Example 2:
x^2 + 18x + 81

Solution:
First find the square root of x^2 y of 81, then open a square bracket.

Then it remains

$\sqrt{x^2} = x$

$\sqrt{81}$ = 9

$(x + 9)^2$

Answer: $(x + 9)^2$

Example 3:
$x^2 + 14x + 49$
Solution:
First find the square root of x^2 and of 49, then open a square bracket.
Then it is
$\sqrt{x^2} = x$

$\sqrt{49} = 7$

$(x + 7)^2$
 Answer: $(x + 7)^2$

Example 4:
$x^2 + 16x + 64$
Solution:
First find the square root of x^2 and of 64, then open a square bracket.
Then
it is $\sqrt{x^2} = x\sqrt{64} = 8$

$(x + 8)^2$
 Answer: $(x + 8)^2$

Example 5:
$x^2 + 24x + 144$

Solution:
First find the square root of x^2 y of 144, then open a square bracket.

Then **it remains** $\sqrt{x^2} = x$
$\sqrt{144} = 12$

$(x + 12)^2$
 Answer: $(x + 12)^2$

14.6 Trinomial of the form ax² + bx + c

Examples of a trinomial of the form ax² + bx + c

Example 1:
5x² + 7x + 2

Solution:

$$\frac{5x^2 + 7x + 2}{5}$$

$$\frac{5(5x^2) + 7(5x) + 5(2)}{5}$$

$$\frac{(5x^2) + 7(5x) + 10}{5}$$

Then it is $\dfrac{(5x + 5)(5x + 2)}{5}$

Answer: (x+ 1) (5x+ 2)

Example 2:
3x² - 10x + 8

Solution:

$$\frac{3x^2 - 10x + 8}{3}$$

$$\frac{3(3x^2) - 10(3x) + 3(8)}{3}$$

$$\frac{(3x^2) - 10(3x) + 24}{3}$$

Then it is $\dfrac{(3x - 6)(3x - 4)}{3}$

Answer: (x - 2) (3x - 4)

Example 3:
4x² + 14x + 10

Solution: $\dfrac{4x^2 + 14x + 10}{4}$ $\dfrac{4(4x^2) + 14(4x) + + 4(10)}{4}$

$$\frac{(4x^2) + 14(4x) + 40}{4}$$

Then it remains $\dfrac{(4x + 10)(4x + 4)}{4}$

Answer: (4x + 10) (x+1)

Example 4:
6x² + 16x + 10

Solution: $\dfrac{6x^2 + 16x + 10}{6}$

$$\frac{6(6x^2) + 16(6x) + 6(10)}{6}$$

$$\frac{(6x^2) + 16(6x) + 60}{6}$$

Then it remains $\dfrac{(6x + 10)(6x + 6)}{6}$

Answer: (6x + 10) (x + 1)

Example 5:
3x² + 13x + 10
Solution:

$$\frac{3x^2 \ + \ 13x \ + \ 10}{3}$$

$$\frac{3(3x^2) \ + \ 13(3x) \ + \ 3(10)}{3}$$

$$\frac{(3x^2) \ + \ 13(3x) \ + \ 30}{3}$$

$$\frac{(3x \ + \ 10)(3x \ + \ 3)}{3}$$

Then it is

Answer: (3x + 10) (x + 1)

14.7 Trinomial of the form x² +bx +c

Examples of trinomial in the form x² + bx + c

Example 1:

x² + 10x + 9

Solution:

First take square root of x²

$$\sqrt{x^2} \ = \ x$$

Then we look for two numbers that when multiplied give the number 9 and added together give the number 10.

Those two numbers are 9 and 1, then two parentheses will be opened.

(x + 9) (x + 1)

Answer: (x + 9) (x + 1)

Example 2:

x² + 6x + 5

Solution:
First take square root of x^2
$$\sqrt{x^2} = x$$

We are looking for two numbers that when multiplied give the number 5 and when added together give the number 6.
These numbers are 5 and 1, then two parentheses will be opened.
(x + 5)(x + 1)
Answer: (x + 5) (x+1)

Example 3:
$x^2 + 2x + 1$

Solution:
First take square root of x^2
$$\sqrt{x^2} = x$$

We look for two numbers that when multiplied give 1 and added together give the number 2.
Those numbers are 1, then two parentheses will be opened.
(x + 1) (x + 1)
Answer: (x + 1) (x + 1)

Example 4:
$x^2 + 3x + 2$

Solution:
First take square root of x^2
$$\sqrt{x^2} = x$$

Then we look for two numbers that when multiplied give the number 2 and when added together give the number 3.
Those numbers are 2 and 1.
(x + 2) (x + 1)
Answer: (x + 2) (x + 1)

Example 5:
$x^2 - 10x + 9$

Solution:
First take square root of x^2
$$\sqrt{x^2} = x$$

Then we look for two numbers that when multiplied give 9 and added together give the number -10.

Those numbers are -9 and -1

(x - 9) (x - 1)

Answer: (x - 9) (x - 1)

Chapter 15: First degree equation.
Definition of first degree equation.

First degree equation.
It is of the form ax + b= 0

Where A is different from 0 or from any other equation in which by transposing terms and simplifying adopt that expression.

Steps to solve a first degree equation.

Step 1:
Parentheses are removed if any.

Step 2:
Remove denominators if any.

Step 3:
We group the x terms in one member and also group the independent terms in another member.

Step 4:
The like terms are reduced.

Step 5:

The unknown of the first degree equation is cleared.

15.2 Exercises with first degree equations

Exercise 1

2x - 10= 0

Solution:

2x= 0 + 10

2x= 10

Then it is $\dfrac{2x}{2} = \dfrac{10}{2}$

X= 5

Answer: x= 5

Check

2x - 10= 0

2(5) - 10=

10 - 10=0

The final conclusion reached is that the answer is correct.

Exercise 2
6x - 3= 0

Solution:

6x= 0 + 3

6x = 3

Then it remains $\dfrac{6x}{6} = \dfrac{3}{6}$

X= $\dfrac{3}{6}$

Simplifying it is that X= $\dfrac{1}{2}$

Answer: X= $\dfrac{1}{2}$

Check

6x - 3= 0

$6(\dfrac{1}{2}) - 3 =$

$\dfrac{6}{2} - 3 =$

Then it remains $\dfrac{6}{2} - \dfrac{3}{1} = \dfrac{6}{2} - \dfrac{6}{2} = \dfrac{0}{2} = 0$

The final conclusion is that the answer is correct.

Exercise 3
4x - 16= 0

Solution:

4x= 0 + 16

4x= 16

Then it remains $\dfrac{4x}{4} = \dfrac{16}{4}$

X= 4

Answer: X= 4
Check:

4x - 16= 0

4(4) - 16=

That leaves 16 - 16= 0

Exercise 4
50x - 10= 0

Solution:

50x= 0 + 10

50x= 10

Then it remains $\dfrac{50x}{50} = \dfrac{10}{50}$

X= $\dfrac{10}{50}$

Simplifying, X= $\dfrac{1}{5}$

Answer: X= $\dfrac{1}{5}$

Verification:

50x - 10= 0

$50(\dfrac{1}{5})$ – 10 =

$\dfrac{50}{5}$ – 10 =

Then it remains $\dfrac{50}{5} - \dfrac{10}{1} = \dfrac{50}{5} - \dfrac{50}{5} = 0$

Exercise 5
10x + 20= 0

Solution:

10x= 0 - 20

10x= -20

Then there is $\dfrac{10x}{10} = \dfrac{-20}{10}$

X= -2
Answer: X= -2
Verification:
10x + 20= 0
10(-2) + 20=
0

Then -20 + 20= 0

Exercise 6
2 (2x - 5)= 30
Solution:
2 × 2x= 4x
2 × -5= -10
4x - 10= 30
4x= 30 + 10
4x= 40

Then remains. $\dfrac{4x}{4} = \dfrac{40}{4}$

x= 10
Answer: x= 10
Verification:
4x - 10= 30
4(10) - 10=
Then 40 - 10= 30
Exercise 7
5x + 5= -10x
Solution:

5x + 10x= 0 - 5
15x= -5

$$\frac{15x}{15} = \frac{-5}{15}$$

Then x= $\frac{-5}{15}$

Simplifying, x= $\frac{-1}{3}$

Answer: x= $\frac{-1}{3}$

Exercise 8
3x - 15= 3
Solution:
3x= 3 + 15
3x= 18

Then there remains $\frac{3x}{3} = \frac{18}{3}$
x= 6
Answer: x= 6
Verification:
3x - 15=3
3(6) - 15= Then
18 - 15= 3

Exercise 9

$$\frac{1}{2}x + \frac{3}{5} = \frac{3}{4}x - \frac{2}{3}$$

Solution:

$$\frac{1}{2}x - \frac{3}{4}x = \frac{-2}{3} - \frac{3}{5}$$

Then it is $\dfrac{-1}{4}x = \dfrac{-19}{15}$

$x = \dfrac{\dfrac{-19}{15}}{\dfrac{-1}{4}}$

$x = \dfrac{-76}{-15}$

Answer: x= $\dfrac{76}{15}$

Exercise 10

$2(3y - 1) = \dfrac{5y}{3}$

Solution:

$6y - 2 = \dfrac{5y}{3}$

Then the fraction has to be eliminated as follows:

$3(6y) + 3(-2) = \dfrac{3(5y)}{3}$

Then it is

$18y - 6 = \dfrac{15y}{3}$

$18y - 6 = 5y$

$18y - 5y = 6$

$13y = 6$

Then there is $\dfrac{13y}{13} = \dfrac{6}{13}$

Answer: Y= $\dfrac{6}{13}$

15.3 Problems with first degree equations

Problem 1.
Its double plus 5 equals 51
Solution:
The first degree equation is stated in the following form:
2x + 5= 51
2x= 51 - 5
2x= 46.

Then it is $\dfrac{2x}{2} = \dfrac{46}{2}$

x= 23
Answer: 23

Problem 2.
Find two consecutive
numbers whose sum is 51
Solution:
As the first degree equation is posed it is in the following form:
x + x + x + 1= 51
x + x= 51 - 1
2x= 50.

It remains $\dfrac{2x}{2} = \dfrac{50}{2}$ **x= 25**

Then the following is done:
25 + 1= 26
Answer: The numbers are 25 and 26.

Problem 3.
Find 3 consecutive numbers whose sum will be 219
Solution:
The first degree equation will be stated as follows:
x + x + x + 1 + x + 2= 219
x + x + x= 219 - 2 - 1
3x= 216.

It remains $\dfrac{3x}{3} = \dfrac{216}{3}$

x= 72
Then we do the following:
72 + 1= 73
72 + 2= 74
Answer: The numbers are 72, 73 and 74.

PA

Problem 4.
How long is a rope if its third quarter measures 200m?
Solution:
The posed first degree equation is left in the following form:
$\dfrac{3x}{4} = 200$ $x= \dfrac{4 \times 200}{3}$

$x= \dfrac{800}{3}$
x= 266.66m

Answer: The rope measures 266.66m.

Problem 5.
5 years ago Peter's age was three times that of his sister Mary who is 5 years old. How many years must pass before Mary is Peter's age today?
Solution:
The first degree equation posed will be as follows:
x - 5= 3 × 10
x - 5= 30
x= 30 + 5
x= 35
Then
35 - 15= 20
Answer: Peter is 35 years old and Mary will be 35 years old in 20 years.

Chapter 16: Full quadratic equation.

Complete quadratic equation by the factorization method.

Examples of complete quadratic equation by the factorization method.
Example 1:
$x^2 + 2x + 1 = 0$

Solution:
First we analyze if the complete quadratic equation is ordered in descending order. The term with x squared goes, then the term with the variable x raised to the exponent and finally the independent term.
 The complete quadratic equation has to be equal to 0
$(x + 1)(x + 1) = 0$
$x + 1 = 0$
$x = 0 - 1$
$x = -1$
$x_1 = -1$

$x + 1 = 0$
$x = 0 - 1$
$x = -1$
$x_2 = -1$
Answer: Are $x_1 = -1$ and $x_2 = -1$
Check
$x^2 + 2x + 1$

Substitute the value of x1 obtained in the full quadratic equation.

$1(-1)^2 + 2(-1) + 1 =$

$1 - 2 + 1 = 0$

Check 1 is correct.

Check 2

$x^2 + 2x + 1 = 0$

The value of x2 is substituted into the original full quadratic equation.

$1(-1)^2 + 2(-1) + 1 =$

$1 - 2 + 1 = 0$

Check 2 is correct.

Example 2:

$x^2 + 6x + 5 = 0.$

Solution:

$(x + 5)(x + 1) = 0$

$x + 5 = 0$

$X = 0 - 5$

$X = -5$

$X1 = -5$

$x + 1 = 0$

$X = 0 - 1$

$X = -1$

$X2 = -1$

Answer: The solutions are x1=-5 and x2=-1

Check

$x^2 + 6x + 5 = 0$

Substitute the value of x1 into the original quadratic equation.

$1(-5)^2 + 6(-5) + 5 =$

$25 - 30 + 5 =$

$-5 + 5 = 0$

The check is correct.

Check 2

$x^2 + 6x + 5 = 0$ **The**

value of x2 is substituted into the original full quadratic equation.

This leaves

$1(-1)^2 + 6(-1) + 5 =$

$1 - 6 + 5 =$

$-5 + 5 = 0$

Check 2 is correct.

Example 3:

$x^2 + 7x + 10 = 0$

Solution:
$(x + 5)(x + 2) = 0$
$X + 5 = 0$
$X = 0 - 5$
$X = -5$
$X1 = -5$
$X + 2 = 0$
$X = 0 - 2$
$X = -2$
$X2 = -2$
Answer: The solutions are x1=-5 and x2=-2
Check
$x^2 + 7x + 10 = 0$
Substitute the value of x1 into the original quadratic equation.
Then it is
$1(-5)^2 + 7(-5) + 10 =$
$25 - 35 + 10 =$
$-10 + 10 = 0$
The check is correct.
Check 2
$x^2 + 7x + 10 = 0$
The value of x2 is substituted into the original full quadratic equation.
Then it is
$1(-2)^2 + 7(-2) + 10 =$
$4 - 14 + 10$
$-10 + 10 = 0$
Check 2 is correct.

Example 4:
$x^2 + 8x + 12 = 0.$
Solution:
$(x + 6)(x + 2) = 0$
$x + 6 = 0$
$x = 0 - 6$
$x = -6$
$x1 = -6$
$x + 2 = 0$
$x = 0 - 2$
$x = -2$
Then $X2 = -2$
Answer: The solutions are X1= -6 and X2= -2
Check
$x^2 + 8x + 12 = 0$

Substitute the value of x1 into the original quadratic equation.
Then it is
1(-6)² + 8(-6) + 12=
36 - 48 + 12=
-12 + 12= 0
The check is correct.
Verification 2
$x^2 + 8x + 12 = 0$
We substitute the value of x2 into the original quadratic equation.
Then it is
1(-2)² + 8(-2) + 12=
4 - 16 + 12=
-12 + 12=0 Check 2 is correct.

16.2 Complete quadratic equation by the general formula method.

Examples of quadratic equation by the general formula method.
Example 1:
$x^2 + 3x + 2 = 0$

Solution:
This equation is of the following form:
$ax^2 + bx + c = 0$
Then A equals 1, B equals 3 and C equals 2.

$$x = \frac{-b \pm \sqrt{b^2 - 4(a)(c)}}{2a}$$

$$x = \frac{-3 \pm \sqrt{3^2 - 4(1)(2)}}{2}$$

$$X = \frac{-3 \pm \sqrt{9 - 8}}{2}$$

$$X = \frac{-3 \pm \sqrt{1}}{2}$$

$$X = \frac{-3 \pm 1}{2}$$

$$X = \frac{-3 - 1}{2}$$

$$X = \frac{-3 + 1}{2} \quad X = \frac{-4}{2}$$

$$X= \frac{-2}{2}$$

X1= -2

X2= -1

Answer: The solutions are x1=-2 and x2=-1

Check

$x^2 + 3x + 2=0$

Substitute the value of x1 into the original quadratic equation.
Then it is

$1(-2)^2 + 3(-2) + 2=$

$4 - 6 + 2=$

$-2 + 2=0$

The check is correct.

Verification 2

$x^2 + 3x + 2= 0$

We substitute the value of x2 into the original quadratic equation.
Then it is

$1(-1)^2 + 3(-1) + 2=$

$1 - 3 + 2=$

$-2 + 2= 0$

Check 2 is correct.

Example 2:

$x^2 + 10x + 9= 0$

Solution:

$ax^2 + bx + c= 0.$

A=1, B=10, C=9

Then it is

$$x= \frac{-b \pm \sqrt{b^2 - 4(a)(c)}}{2a} \qquad x= \frac{-10 \pm \sqrt{10^2 - 4(1)(9)}}{2} \qquad x=$$

$$x= \frac{-10 \pm \sqrt{100 - 36}}{2}$$

$$x= \frac{-10 \pm \sqrt{64}}{2}$$

$$x= \frac{-10 \pm 8}{2}$$

$$x= \frac{-10 - 8}{2}$$

$$x = \frac{-10 + 8}{2}$$

$$x = \frac{-18}{2}$$

$$x = \frac{-2}{2}$$

$x1 = -9$

$x2 = -1$

Answer: The solutions are x1=-9 and x2=-1

Check

$x^2 + 10x + 9 = 0$

The value of x is substituted into the original quadratic equation.
This leaves

$1(-9)^2 + 10(-9) + 9 =$

$81 - 90 + 9 =$

$-9 + 9 = 0$

The check is correct.

Verification 2

$x^2 + 10x + 9 = 0$

The value of x2 is substituted into the full quadratic
equation.
Then it is

$1(-1)^2 + 10(-1) + 9 =$

$1 - 10 + 9 =$

$-9 + 9 = 0$

The second check is correct.

Example 3:

$x^2 + 6x + 5 = 0$

Solution:

$ax^2 + bx + c = 0$

A=1 , B=6 , C=5

Then there remains

$$x = \frac{-b \pm \sqrt{b^2 - 4(a)(c)}}{2a}$$

$$x = \frac{-6 \pm \sqrt{6^2 - 4(1)(5)}}{2}$$

$$x = \frac{-6 \pm \sqrt{36 - 20}}{2}$$

$$x = \frac{-6 \pm \sqrt{16}}{2}$$

$$x = \frac{-6 \pm 4}{2}$$

$$x = \frac{-6 - 4}{2}$$

$$x = \frac{-6 + 4}{2}$$

$$x = \frac{-10}{2}$$

$$x = \frac{-2}{2}$$

$x_1 = -5$

$x_2 = -1$

Answer: The solutions are $x_1 = -5$ and $x_2 = -1$

Verification

$x^2 + 6x + 5 = 0$

Substitute the value of x1 into the original quadratic equation

Then it is.

$1(-5)^2 + 6(-5) + 5 =$

$25 - 30 + 5 =$

$-5 + 5 = 0$

The check is correct.

Check 2

$x^2 + 6x + 5 = 0$

We substitute the value of x2 into the original full quadratic equation.

This leaves

$1(-1)^2 + 6(-1) + 5 =$

$1 - 6 + 5 =$

$-5 + 5 = 0$

Check 2 is correct.

Example 4:

$x^2 + 8x + 7 = 0$

Solution:

$ax^2 + bx + c = 0$

A=1, B=8, C=7

Then there remains

$$x = \frac{-b \pm \sqrt{b^2 - 4(a)(c)}}{2a}$$

$$x = \frac{-8 \pm \sqrt{8^2 - 4(1)(7)}}{2} \qquad x = \frac{-8 \pm \sqrt{64 - 28}}{2}$$

$$x = \frac{-8 \pm \sqrt{36}}{2}$$

$$x = \frac{-8 \pm 6}{2}$$

$$x = \frac{-8 - 6}{2}$$

$$x = \frac{-8 + 6}{2}$$

$$x = \frac{-14}{2}$$

$$x = \frac{-2}{2}$$

x1= -7

x2= -1

Answer: The solutions are x1=-7 and x2=-1

Verification

$x^2 + 8x + 7 = 0$

The value of x1 is substituted into the original quadratic equation. Then it is

$1(-7)^2 + 8(-7) + 7 =$

49 - 56 + 7=

-7 + 7= 0

The check is correct.

Check 2

$X^2 + 8x + 7 = 0$

The value of x2 is substituted into the original quadratic equation.

Then it remains.

$1(-1)^2 + 8(-1) + 7 =.$

1 - 8 + 7=

-7 + 7= 0

Check 2 is correct.

Example 5:
$x^2 + 9x + 8= 0$

Solution:
$ax^2 + bx + c= 0$
A=1, B=9, C= 8
Then there is

$$x= \frac{-b \pm \sqrt{b^2 - 4(a)(c)}}{2a}$$

$$x= \frac{-9 \pm \sqrt{9^2 - 4(1)(8)}}{2}$$

$$x= \frac{-9 \pm \sqrt{81 - 32}}{2}$$

$$x= \frac{-9 \pm \sqrt{49}}{2}$$

$$x= \frac{-9 - 7}{2}$$

$$x= \frac{-9 + 7}{2}$$

$$x= \frac{-16}{2}$$

$$x= \frac{-2}{2}$$

x1= -8
x2= -1
Answer: The solutions are x1=-8 and x2=-1
Verification
$x^2 + 9x + 8= 0$
The value of x1 is substituted into the original quadratic equation.
Then it is
$1(-8)^2 + 9(-8) + 8=$
64 - 72 + 8=
-8 + 8= 0
The check is correct.
Check 2
$x^2 + 9x + 8= 0$

The value of x2 is substituted into the original quadratic equation.

$1(-1)^2 + 9(-1) + 8=$

$1 - 9 + 8=$

$-8 + 8= 0$

Check 2 is correct.

16.3 Complete quadratic equation by the method of completing the squared trinomial

Examples of quadratic
equation by the method of completing the square trinomial.
Example 1:
$x^2 + 2x - 1= 0$

Solution:
$x^2 + 2x= 1$
Then it is

$x^2 + 2x + \left(\dfrac{2}{2}\right)^2 = 1 + \left(\dfrac{2}{2}\right)^2$

$(x + 1)^2 = 1 + (1)^2$

$(x + 1)^2 = 1 + 1$

$(x + 1)^2 = 2$

$x + 1= \sqrt{2}$

$x= -1 + \sqrt{2}$

$x= -1 - \sqrt{2}$

$x= -1 + \sqrt{2}$ $x1= -1 - \sqrt{2}$ $x2= -1 + \sqrt{2}$

Example 2:
$x^2 + 2x - 4= 0$
Solution:
$x^2 + 2x= 4$
This leaves

$x^2 + 2x + \left(\dfrac{2}{2}\right)^2 = 4 + \left(\dfrac{2}{2}\right)^2$

$(x + 1)^2 = 4 + (1)^2$

$(x + 1)^2 = 4 + 1$

$x + 1= \sqrt{5}$

$x= -1 +- \sqrt{5}$

$x= -1 - \sqrt{5}$

$x= -1 + \sqrt{5}$

$x1= -1 - \sqrt{5}$ $x2= -1 + \sqrt{5}$

Example 3:

$X^2 + 6x - 3 = 0$

Solution:

$X^2 + 6x = 3$

Then it remains

$X^2 + 6x + (\frac{6}{2})^2 = 3 + (\frac{6}{2})^2$

$(x + 3)^2 = 3 + (3)^2$

$(x + 3)^2 = 3 + 9$

$X + 3 = \sqrt{12}$

$x = -3 + \sqrt{12}$

$x = -3 - \sqrt{12}$

$x = -3 + \sqrt{12}$ $x1 = -3 - \sqrt{12}$ $x2 = -3 + \sqrt{12}$

Example 4:
$x^2 + 10x - 5 = 0$

Solution:

$x^2 + 10x = 5$

Then it remains

$X^2 + 10x + (\frac{10}{2})^2 = 5 + (\frac{10}{2})^2$

$(x + 5)^2 = 5 + (5)^2$

$(x + 5)^2 = 5 + 25$

$x + 5 = \sqrt{30}$

$x = -5 +- \sqrt{30}$

$x = -5 - \sqrt{30}$

$x = -5 + \sqrt{30}$

$X1 = -5 - \sqrt{30}$

$X2 = -5 + \sqrt{30}$

Example 5:
$x^2 + 4x - 2 = 0$

Solution:

$X^2 + 4x = 2$

$x^2 + 4x + (\frac{4}{2})^2 = 2 + (\frac{4}{2})^2$

$(x + 2)^2 = 2 + (2)^2$

$(x + 2)^2 = 2 + 4$

$x + 2 = \sqrt{6}$

$x = -2 +- \sqrt{6}$

$X = -2 - \sqrt{6}$

$x = -2 + \sqrt{6}$

Then

$x_1 = -2 - \sqrt{6}$

$X_2 = -2 + \sqrt{6}$

16.4 Exercises of complete quadratic equation.

Exercise 1

$X^2 + 10x + 9 = 0$

Solution:

$(x + 9)(x + 1) = 0$

$X + 9 = 0$

$X = 0 - 1$

$X = -9$

$X + 1 = 0$

$X = 0 - 1$

$X = -1$

Then it remains

$X1 = -9$
$x2 = -1$

Answer: The solutions are x1=-9 and x2=-1

Exercise 2

$X^2 + 15x + 50 = 0$

Solution:

$(x + 10)(x + 5) = 0$

$X + 10 = 0$

$X = 0 - 10$

$X = -10$

$X + 5 = 0$

$X = 0 - 5$

$X = -5$

Then it remains

X1=-10

X2= -5

Answer: The solutions are x1=-10 and x2=-5.

Exercise 3
$x^2 + 9x + 14 = 0$

Solution:
$(x + 7)(x + 2) = 0$
x + 7= 0
x= 0 - 7
x= -7
x + 2= 0
x= 0 - 2
x=-2
Then it is
x1= -7
x2= -2
Answer: The solutions are x1=-7 and x2=-2

Exercise 4
$x^2 + 5x + 6 = 0$
Solution:
$(x + 3)(x + 2) = 0$
x + 3=
0
x= 0 - 3
x= -3
x + 2= 0
x= 0 - 2
x= -2

Answer: The solutions are x1=-3 and x2= -2.

Exercise 5
$x^2 - 2x + 1 = 0$
Solution:
$(x - 1)(x - 1)(x - 1) = 0$
x - 1=
0

x= 0 + 1
x= 1
x - 1= 0
x= 0 + 1
x= 1
Then it is
x1= 1
x2= 1
Answer: The solutions are x1=1 and x2=1

Chapter 17: Pure Incomplete Quadratic Equation.
Definition of pure incomplete quadratic equation.

Pure incomplete quadratic equation.

It is called a pure incomplete quadratic equation because it is of the following form
$ax^2 + c = 0$

Steps to solve a pure incomplete quadratic equation.
Step 1:
First, the independent term is passed to the second member of the equation by changing the sign.
Step 2:
You pass the term a to the second member by dividing.
Step 3:
Square root is taken in both members of the equation, two solutions will be obtained one solution with positive and the other solution with negative.

Example:
$3x^2 - 75 = 0$
Solution:
$ax^2 + c = 0$
Then we have
$3x^2 = 25$
$$x^2 = \frac{75}{3}$$
x^2
$= 25 \; \sqrt{x^2} = \sqrt{25}$
$x = 5$
$x = -5$
Then we are left with
$x_1 = 5$
$x_2 = -5$
Answer: The solutions are x1=5 and x2=-5

17.2 Exercises of pure incomplete quadratic equation.

Exercise 1
$2x^2 - 8 = 0$
Solution:
$2x^2 = 8$
$x^2 = \dfrac{8}{2}$
$x^2 = 4 \quad \sqrt{x^2} = \sqrt{4}$
$x = 2$
$x = -2$

Answer: The solutions are x1=2 and x2= -2.

Exercise 2
$4x^2 - 100 = 0$
Solution:
$4x^2 = 100$
$x^2 = \dfrac{100}{4}$
$x^2 = 25 \quad \sqrt{x^2} = \sqrt{25}$
$x = 5$
$x = -5$

Answer: The solutions are x1=5 and x2=-5.

Exercise 3
$9x^2 - 81 = 0$
Solution:
$9x^2 = 81$
$x^2 = \dfrac{81}{9}$
$x^2 = 9 \quad \sqrt{x^2} = \sqrt{9}$
$x = 3$
$x = -3$
Then
$x1 = 3$

x2= -3
Answer: The solutions are x1=3 and x2=-3

Exercise 4
$4x^2 - 16 = 0$
Solution:
$4x^2 = 16$
$x = {}^2\dfrac{16}{4}$
$x^2 = 4$
$\sqrt{x^2} = \sqrt{4}$
x= 2
x= -2
Then it is
x1= 2
x2= -2
Answer: The solutions are x1=2 and x2=-2

Exercise 5
$x^2 - 36 = 0$
Solution:
$x^2 = 36$
$x^2 = \dfrac{36}{1}$
x^2
$= 36 \quad \sqrt{x^2} = \sqrt{36}$
x= 6
x= -6
Then
x1= 6
x2=-6
Answer: The solutions are x1=6 and x2=-6

Mixed incomplete quadratic equation.
A mixed incomplete quadratic equation is a quadratic equation that is of the following form $ax^2 + bx = 0$

Steps to solve a mixed incomplete quadratic equation.
 Step 1:
The common factor is extracted.
 Step 2:
Since you will have a product the product will be equal to 0, one factor is as another factor is 0.
Step 3:

The solutions will be of the following form $x1 = 0$ and $x2 =. \dfrac{-b}{a}$

Example:
$x^2 - 6 = 0$
Solution:
$x(x - 6) = 0$
$x = 0$
$x - 6 = 0$
$x = 0 + 6$
$x = 6$
Then it is
$x1 = 0$
$x2 = 6$
Answer: The solutions are x1=0 and x2= 6

18.2 Incomplete mixed quadratic equation exercises.

Exercise 1
$x^2 + 10x= 0$
Solution:
$x (x + 10)= 0$
$x= 0$
$x + 10=0$
$x= 0- 10$ $x=$
-10
Then
$x1=0$
$x2= -10$
Answer: The solutions are x1=0 and x2=-10

Exercise 2
$x^2 - 4x= 0$
Solution:
$x(x - 4)=0$
$x=0$
$x - 4=0$
$x= 0 + 4$
$x= 4$
Then $x1=0$
$x2=4$
Answer: The solutions are x1=0 and x2= 4

Exercise 3
$x^2 + 2x=0$
Solution:
$x(x + 2)=0$
$x=0$
$x + 2= 0$
$x= 0$ $x= 0 - 2$
$x= -2$
Then the solutions are
$x1=0$
$x2=-2$
Answer: The solutions are x1=0 and x2=-2

Exercise 4
$9x^2 - 15x=0$
Solution:
We take out the common factor 3x
3x 3x(x - 5)= 0
3x=
$$0 \quad \frac{3x}{3} = \frac{0}{3}$$
x= 0
3x - 5= 0
3x= 0 + 5
3x=
$$5 \quad \frac{3x}{3} = \frac{5}{3}$$
x= $\frac{5}{3}$ Then
x1= 0
x2= $\frac{5}{3}$ Answer: The solutions are x1=0 and x2= $\frac{5}{3}$

Exercise 5
$5x^2 - 25x= 0$
Solution:
We take out the common factor 5x
Then we get
5x(x - 5)= 0
5x=
$$0 \quad \frac{5x}{5} = \frac{0}{5}$$
x= 0
5x- 5= 0
5x= 0 + 5
5x=
$$5 \quad \frac{5x}{5} = \frac{5}{5}$$
x= 1
Then it is
x1=0
x2= 1

Answer: The solutions are x1=0 and x2=1

Chapter 19: System of equations with 2 unknowns.
Exercises of system of equations with 2 unknowns by the method of addition and subtraction.

Exercise 1
x + y= 3 (1)
x- y= 1 (2)
Solution:
x + y= 3 (1)
x - y= 1 (2)
The unknown Y will be eliminated from the system
2x= 4
$$\frac{2x}{D\,2} = \frac{4}{2}$$
x= 2
Substitute the value of x in the first equation of the system.
x + y= 3 (1)
1(2) + y= 3
2 + y= 3
y= 3 - 2
y= 1
Answer: The solutions are x= 2, y= 1
Verification
x + y= 3 (1)
1(2) + 1(1)=
2 + 1= 3
The check is correct.
Verification 2
x - y= 1 (2) 1(
2) - 1(1)=
2 - 1= 1
Check 2 is correct.

Exercise 2
x + y= 9 (1)
x - y= 1 (2)

Solution:
The unknown Y will be eliminated from the system.

2x=

$$\frac{2x}{2} = \frac{10}{2}$$

10

X= 5

The value of x is substituted into the first equation of the system.

x + y= 9 (1)
1(5) + y= 9
5 + y= 9
y= 9 - 5
y= 4

Answer: The solutions are x= 5, y= 4

Verification

x + y= 9 (1)
1(5) + 1(4)=
5 + 4= 9

The check is correct.

Verification 2

x - y= 1 (2)
1(5) - 1(4)=
5 - 4= 1

Check 2 is correct.

Exercise 3

x + y= 4 (1)
x - y= 2 (2)

Solution:
We solve this system of equations with 2 unknowns by the method of addition and subtraction as follows:

x + y= 4 (1)
x - y= 2 (2).

The unknown is eliminated from the system of equations.

2x=

$$\frac{2x}{2} = \frac{6}{2}$$

6

X= 3

The value of x is substituted into the first equation of the system

x + y= 4 (1)
1(3) + y= 4
3 + y= 4

y= 4 - 3
y= 1
Answer: The solutions are x= 3
, y=1
Verification
x + y= 4 (1)
1(3) + 1(1)=
3 + 1= 4
The check is correct.
Check 2
x - y= 2 (2)
Then it remains.
1(3) - 1(1)=
3 - 1= 2
Check 2 is correct.

Exercise 4
2x + 4y= 10 (1)
x + 3y= 7 (2)
Solution:
We solve this system of equations with 2 unknowns by the addition and subtraction method as follows:
2x + 4y= 10 (1)
x + 3y= 7 (2).
The first equation will be multiplied by 1 and the second equation by -2
2x + 4y = 10
-2x - 6y= -14
The unknown x will be eliminated from the system of equations.
-2y= -4

$$\frac{-2y}{-2} = \frac{-4}{-2}$$

Y= 2
The value of y is substituted into the first equation of the system.
2x + 4y= 10 (1)
2x + 4(2)= 10
2x + 8= 10
2x= 10 - 8
2x=

$$\frac{2x}{2} = \frac{2}{2}$$

X=1
Answer: The solutions are x=1, y=2

Verification
2x + 4y= 10 (1)
2(
1
) + 4(2)=
2 + 8= 10
The check is correct.
Verification 2
x + 3y= 7 (2)
1(1) + 3(2)=
1 + 6= 7

Check 2 is correct.

Exercise 5
2x + 4y= 10 (1)
2x + 6y= 14 (2)
Solution:
We solve this system of equations with 2 unknowns by the method of addition and subtraction as follows:
2x + 4y= 10 (1)
2x + 6y= 14 (2).
The first equation is multiplied by 2 and the second by -2
4x + 8y= 20
-4x - 12= -28
The unknown x is eliminated from the system of equations.
-4y=

$$\frac{-4y}{-8} \frac{}{-4} = \frac{-8}{-4}$$

Y= 2
The value of (y) is substituted into the second equation of the system.
2x + 6y= 14 (2)
2x + 6(2)= 14
2x + 12= 14
2x= 14 - 12
2x=

$$\frac{2x}{2} \frac{}{2} = \frac{2}{2}$$

X= 1
Answer: x= 1, y= 2
Verification
2x + 4y= 10 (1)

2(1) + 4(2)=
2 + 8=10
The check is correct.
Verification 2
2x + 6y= 14 (2)
2(1) + 6(2)=
2 + 12= 14
Check 2 is correct.

19.2 Exercises of a system of equations with 2 unknowns by the method of equalization.

Exercise 1
2x + 3y= 0 (1)
3x + 4y= 1 (2)
Solution:
How to solve this system of equations with 2 unknowns by the method of equalization as follows:
2x + 3y= 0 (1)
3x + 4y= 1 (2).
We do the following:
3y= 0 - 2x

$$\frac{0 - 2x}{3}$$

and=
4y= 1 - 3x

$$y= \frac{1 - 3x}{4}$$

$$\frac{0 - 2x}{3} = \frac{1 - 3x}{4}$$

4(0) + 4(-2x)= 3(1) + 3(-3x)
0 - 8x= 3 - 9x
-8x + 9x= 3 - 0
X= 3
The value of x is substituted into the second equation of the system.
3x + 4y= 1 (2)
3(3) + 4y= 1
9 + 4y= 1

4y= 1 - 9

4y=

$$\frac{4y}{-8\ 4} = \frac{-8}{4}$$

y= -2

Answer: x=3, y=-2

Verification

2x + 3y= 0 (1)

2(3) + 3(-2)=

6 - 6= 0

The check is correct.

Verification 2

3x + 4y= 1 (2)

3(3) + 4(-2)=

9 - 8= 1

Check 2 is correct.

Exercise 2

3x + 2y= 7 (1)

4x - 3y= -2 (2)

Solution:

We solve this system of equations with 2 unknowns by the method of equalization.

3x + 2y= 7(1)

4x - 3y= -2 (2)

2y= 7 - 3x

$$and= \frac{7\ -\ 3x}{2}$$

-3y= -2 - 4x

$$and= \frac{-2\ -\ 4x}{-3}$$

$$\frac{7\ -\ 3x}{2} = \frac{-2\ -\ 4x}{-3}$$

-3(7) -3(-3x)= 2(-2) + 2(-4x)

-21 + 9x= -4 - 8x

9x + 8x= -4 + 21

17x= 17

$$\frac{17x}{17} = \frac{17}{17}$$

X= 1

Substitute the value of x in the first equation of the system.

3x + 2y= 7 (1)

3(1) + 2y= 7
3 + 2y= 7
2y= 7 - 3
2y=

$$\frac{2y}{4} \frac{2}{2} = \frac{4}{2}$$

Y= 2
Answer: x=1, y=2

Verification
3x + 2y= 7 (1)
3(1) + 2(2)=
3 + 4= 7
The check is correct.
Verification 2
4x - 3y= -2 (2) 4(1
)- 3(2)=
4 - 6= -2
Check 2 is correct.

Exercise 3
2x + y= 6 (1)
4x + 3y= 14 (2)
Solution:
We solve this system of equations with 2 unknowns by the method of equalization.
2x + y= 6 (1)
4x + 3y= 14 (2)
2x= 6 - y

$$x= \frac{6 - y}{2}$$

4x= 14 - 3y

$$x= \frac{14 - 3y}{4}$$

$$\frac{6 - y}{2} = \frac{14 - 3y}{4}$$

4(6) + 4(-y)= 2(14) + 2(-3y)
24 - 4y= 28 - 6y
-4y + 6y= 28- 24
2y= 4

$$\frac{2y}{2} = \frac{4}{2}$$

Y= 2

Substitute the value of (y) into equation 2 of the system.

4x + 3y= 14 (2)

4x + 3(2)= 14

4x + 6= 14

4x= 14 - 6

4x= 8

$$\frac{4x}{4} = \frac{8}{4}$$

X= 2

Answer: x=2, y= 2

Verification

2x + y= 6 (1)

2(2) + 1(2)=

4 + 2= 6

The check is correct.

Verification 2

4x + 3y= 14(2)

4(2) + 3(2)=

8 + 6= 14

Check 2 is correct.

Exercise 4

5x + 2y= 2 (1)

x - 2y= 2 (2

)

Solution:

We solve this system of equations with 2 unknowns by the method of equalization.

5x + 2y= 2 (1)

x - 2y= 2 (2)

5x= 2 - 2y

$$x= \frac{2 - 2y}{5}$$

x= 2 + 2y

$$x= \frac{2 + 2y}{1}$$

$$\frac{2 - 2y}{5} = \frac{2 + 2y}{1}$$

$1(2) + 1(-2y)= 5(2) + 5(2y)$

$2 - 2y= 10 + 10y$

$-2y - 10y= 10 - 2$

$-12y= 8$

$$\frac{-12y}{-12} = \frac{8}{-12}$$

$y= .\dfrac{8}{12}$

$y= \dfrac{-2}{3}$

Substitute the value of (y) into the first equation of the system.

$5x + 2y= 2 \ (1)$

$5x + 2(\dfrac{-2}{3} = 2$

$5x - \dfrac{4}{3} = 2$

$5x= 2 + \dfrac{4}{3}$

$5x= \dfrac{10}{3}$

$$\frac{5x}{5} = \frac{\frac{10}{3}}{5}$$

$\dfrac{10}{1} \quad x= \dfrac{10}{15}$

$x= \dfrac{2}{3}$

Answer: $x= \dfrac{2}{3}$, $y= \dfrac{-2}{3}$

Verification

$5x + 2y= 2 \ (1) \ 5$

$(\dfrac{2}{3}) + 2\left(\dfrac{-2}{3}\right) = 0$

$\dfrac{10}{3} - \dfrac{4}{3} = \dfrac{6}{3} = 2$

The check is correct.
Verification 2
x - 2y= 2 (2)

$$1(\frac{2}{3}) - 2(\frac{-2}{3}) =$$

$$\frac{2}{3} + \frac{4}{3} = \frac{6}{3} = 2$$

Check 2 is correct.

Exercise 5
4x + 6y= 2 (1)
6x + 5y= 1 (2)
Solution:
We solve this system of equations with 2 unknowns by the method of equalization.
4x + 6y= 2 (1)
6x + 5y= 1 (2)
6y= 2 - 4x

$$y= \frac{2 - 4x}{6}$$

5y= 1 - 6x

$$y= \frac{1 - 6x}{5}$$

$$\frac{2 - 4x}{6} = \frac{1 - 6x}{5}$$

5(2)+ 5(-4x)= 6(1) + 6(-6x)
10 - 20x= 6 - 36x
-20x + 36x= 6- 10
16x= -4

$$\frac{16x}{16} = \frac{-4}{16}$$

$$x= \frac{-4}{16}$$

$$x= \frac{-1}{4}$$

Substitute the value of x into the first equation of the system.
4x + 6y= 2 (1)

$$4(\frac{-1}{4}) + 6y = 2$$

$$\frac{-4}{4} + 6y = 2$$

$$6y = 2 + \frac{4}{4}$$

$$6y =$$

$$\frac{6y}{3\ 6} = \frac{3}{6}$$

$$y = \frac{3}{6}$$

$$y = \frac{1}{2}$$

Answer: x= $\frac{-1}{4}$ **, y=** $\frac{1}{2}$

Verification

4x + 6y= 2 (1)

$$4(\frac{-1}{4}) + 6\left(\frac{1}{2}\right) =$$

$$\frac{-4}{4} + \frac{6}{2} = \frac{8}{4} = 2$$

The check is correct.

Verification 2

6x + 5y= 1 (2)

$$6(\frac{-1}{4}) + 5\left(\frac{1}{2}\right) =$$

$$\frac{-6}{4} + \frac{5}{2} = \frac{4}{4} = 1$$

Check 2 is correct.

19.3 Exercises of a system of equations with 2 unknowns by the substitution method.

Exercise 1
5x + 2y= 10 (1)
x - 2y= 2 (2)
Solution:
We solve this system of equations with 2 unknowns by the substitution method.
5x + 2y= 10 (1)
x - 2y= 2 (2)
x= 2 + 2y

$$x= \frac{2 + 2y}{1}$$

$$2y + 5(\frac{2 + 2y}{1}) = 10$$

$$2y + \frac{10 + 10y}{1} = 10$$

$$1(2y + \frac{10 + 10y}{1} = 10)$$

2y + 10 + 10y= 10
2y + 10y= 10 - 10
12y= 0

$$\frac{12y}{12} = \frac{0}{12}$$
Y= 0

Substitute the value of (y) in the first equation of the system.
5x + 2y= 10 (1)
5x + 2(0)= 10
5x + 0= 10

5x= 10 - 0

5x=

$$\frac{5x}{5} = \frac{10}{5}$$

X= 2

Answer: x=2, y= 0

Verification

5x + 2y= 10 (1)

5(2) + 2(0)=

10 + 0= 10

The check is correct.

Verification 2

x - 2y= 2 (2)

1(2) - 2(0)=

2 - 0= 2

Check 2 is correct.

Exercise 2

-2x + 4y= -12 (1)

5x - y= 3 (2)

Solution:

We solve this system of equations with 2 unknowns by the substitution method.

-2x + 4y= -12 (1)

5x - y= 3 (2)

4y= -12 + 2x

$$y= \frac{-12 + 2x}{4}$$

$$5x - 1\left(\frac{-12 + 2x}{4}\right) = 3$$

$$5x + \frac{12 - 2x}{4} = 3$$

$$4(5x + \frac{12 - 2x}{4} = 3)$$

20x + 12 - 2x= 12

20x - 2x= 12 - 12

18x=

$$\frac{18x}{18} = \frac{0}{18}$$

X= 0

The value of x is substituted into the second equation of the system.

5x- y= 3 (2) 5

$(0) - y= 3$

$0 - y= 3$

$-y= 3 - 0$

$-y=$

$$\frac{-y}{3 \quad -1} = \frac{3}{-1}$$

$Y= -3$

Answer: x= 0, y=-3 Proof

$-2x + 4y= -12$ (1)

$-2(0) + 4(-3)=$

$-0 - 12= -12$

The check is correct.

Verification 2

$5x- y= 3(2)$

$5(0) - 1(3)=$

$0 - 3= -3$

Check 2 is correct.

Exercise 3

$4x+ 5y= 15$ (1)

$6x - 5y= 5$ (2)

Solution:

We solve this system of equations with 2 unknowns by the substitution method.

$4x + 5y= 15$ (1)

$6x- 5y= 5$ (2)

$5y= 15 - 4x$

$$\text{and} = \frac{15 - 4x}{5}$$

$$6x - 5(\frac{15 - 4x}{5}) = 5$$

$$6x - \frac{75 + 20x}{5} = 5$$

$$5(6x - \frac{75 + 20x}{5} = 5)$$

$30x - 75 + 20x= 25$

$30x+ 20x= 75 + 25$

$50x=$

$$\frac{50x}{100 \quad 50} = \frac{100}{50}$$

X= 2

Substitute the value of x in the first equation of the system.

4x + 5y= 15 (1)

4(2) + 5y= 15

8 + 5y= 15

5y= 15 - 8

5y= 7

$$\frac{5y}{5} = \frac{7}{5} \quad Y= \frac{7}{5}$$

Answer: x= 2, y= $\frac{7}{5}$

Check

4x +5y=15 (1)

$$4(2) + 5(\frac{7}{5}) =$$

$$8 + \frac{35}{5} =$$

$$\frac{8}{1} + \frac{35}{5} = \frac{40}{5} + \frac{35}{5} =$$

$$\frac{40}{5} + \frac{35}{5} =$$

8 + 7= 15

The check is correct.

Verification 2

6x - 5y= 5 (2)

6(2) - 5

$$(\frac{7}{5}) =$$

$$12 - \frac{35}{5} =$$

$$\frac{12}{1} - \frac{35}{5} = \frac{60}{5} - \frac{35}{5} =$$

$$\frac{60}{5} - \frac{35}{5} = \frac{25}{5} = 5$$

Check 2 is correct.

Exercise 4

2x + 3y= -1 (1)
3x + 4y= 0 (2)

Solution:

We solve this system of equations with 2 unknowns by the substitution method.

2x + 3y=-1 (1)
3x + 4y= 0 (2)

4y= 0- 3x

$$\text{and} = \frac{0 - 3x}{4}$$

$$2x + 3(\frac{0 - 3x}{4}) = -1$$

$$2x + \frac{0 - 9x}{4} = -1$$

$$4(2x + \frac{0 - 9x}{4} = -1)$$

8x + 0 - 9x= -4

8x- 9x= -4 - 0

-x = -4

$$\frac{-x}{-1} = \frac{-4}{-1}$$

X= 4

Substitute the value of x in the equation 2

3x + 4y= 0 (2)
3(4) + 4y= 0
12 + 4y= 0
4y= 0 - 12
4y=

$$\frac{4y}{4} = \frac{-12}{4}$$

-12

Y= -3

Answer: X=4, Y=-3

Verification

2x +3y= -1(1)
2(4) + 3(-3)=
8 - 9= -1

The check is correct.

Verification 2

3x + 4y= 0 (2)
3(4)+ 4(-3)=

12 - 12=0
Check 2 is correct.

PA

19.4 Exercises of a system of equations with 2 unknowns by Cramer's method.

Exercise 1
2x + 5y= 4 (1)
x - 5y= 2 (2)
Solution:

We solve this system of equations with 2 unknowns by Cramer's method as follows:

First we find DS
2 × -5= -10
1 × 5= 5
-10- 5= -15
DS= -15
We find the value of x

$$x= \frac{DX}{DS}$$

4 + 5
2 - 5
4 × -5= -20
2 × 5= 10
DX= -20 - 10
DX= -30

$$x= \frac{-30}{-15}$$

X= 2
Then we find DY
2 + 4
1 + 2
2 × 2= 4
1 × 4= 4
Then
DY= 4 - 4
DY= 0
The value of (y) is found as follows:

$$y= \frac{DY}{DS}$$

$$y= \frac{0}{-15}$$

Y= 0
Answer: x= 2, y= 0
Verification
2x + 5y= 4 (1)
2(2) + 5(0)=
4 +0= 4
The check is correct.
Verification 2

x - 5y= 2 (2)
1(2)- 5(0)=
2 - 0= 2
Check 2 is correct.

Exercise 2
2x + 3y= -1 (1)
3x + 4y= 0 (2)
Solution:
We solve this system of equations with 2 unknowns by Cramer's method as follows:
First, DS
2 + 3
3 + 4
2 × 4= 8
3 × 3= 9
DS= 8 - 9
DS= -1
Dx is found as follows:
-1 + 3
0 + 4
-1 × 4= -4
0× 3= 0
DX= -4 - 0
DX= -4
Then we find the value of x
$$x= \frac{DX}{DS}$$
$$x= \frac{-4}{-1}$$
x= 4
Then we find DY
2 - 1
3 + 0
2 × 0= 0
3 × -1= -3
Dy= 0 + 3
DY = 3
We find the value of Y
$$y= \frac{3}{-1}$$
Y= -3

Answer: x= 4, y= -3
Check
2x + 3y= -1 (1)
2(4) + 3
(-3)=
8 - 9= -1
The check is correct.
Verification 2
3x + 4y= 0 (2)
3(4) + 4(-3)=
12 - 12= 0
Check 2 is correct.

Exercise 3
x + y= 2 (1)
2x + y= 5 (2)
Solution:
We solve this system of equations with 2 unknowns by Cramer's method as follows:
First find DS
1 + 1
2 + 1
1 × 1= 1
2 × 1= 2
Ds= 1 - 2
Ds= -1
DX is found in the following way:
2 + 1
5 + 1
2 × 1= 2
5 × 1= 5
DX= 2 - 5
DX= -3
We find the value of x

$$x= \frac{DX}{DS}$$

$$x= \frac{-3}{-1}$$

X= 3
DY is found
1 + 2 2
+ 5

1 × 5= 5
2 × 2= 4
DY=5- 4
DY= 1
Then we find the value of Y

$$y= \dfrac{DY}{DS}$$

$$y= \dfrac{1}{-1}$$

Y= -1

Answer: x= 3, y= -1

Verification

x + y= 2(1)

1(3) + 1(-1)=

3 - 1= 2

The check is correct.

Verification 2

2x + y= 5 (2)

2(3) + 1(-1)=

6 - 1= 5

Check 2 is correct.

Chapter 20: System of equations with 3 unknowns.
Exercises of system of equations with 3 unknowns by the method of addition and subtraction.

Exercise 1
3x + 2y + z=1(1)
5x+3y + 4z=2 (2)
x + y - z= 1 (3)
Solution:
We solve this system of equations with 3 unknowns by the method of addition and subtraction as follows:
We will take the first equation and the third one we will eliminate the unknown z
3x + 2y + z= 1 (1)
x + y - z= 1 (3)
4x + 3y= 2 (4)
Equations 2 and 3 of the system will be taken and the unknown z
5x + 3y + 4z= 2 (2)
x + y - z= 1 (3)
Equation 3 will be multiplied by 4
4x + 4y - 4z= 4
5x + 3y + 4z= 2
4x + 4y - 4z= 4
9x + 7y= 6 (5)
Afterwards we will work with the equation 4 and 5
4x + 3y= 2 (4)
9x + 7y= 6 (5)
Equation 4 is multiplied by 9 and equation 5 by -4
36x + 27y= 18
-36x - 28y= -24
-y=

$$\frac{-y}{-6} = \frac{-6}{-1}$$
-6   - 1   - 1

Y= 6
Substitute the value of (y) in equation 5
9x + 7y= 6 (5)
9x + 7(6)= 6
9x + 42= 6
9x= 6 - 42
9x=

$$\frac{9x}{9} = \frac{-36}{9}$$

-36

x= -4

Substitute the value of x,y in the equation 2

5x + 3y + 4z= 2 (2)

5(-4) + 3(6) + 4z= 2

-20 +18 + 4z= 2

4z= 2 + 20- 18

4z=

$$\frac{4z}{4} = \frac{4}{4}$$

4

Z= 1

Answer: x= -4, y= 6 and z= 1

Check 1

3x+ 2y + z= 1(1)

3(-4) + 2(6)+ 1(1)=

-12 + 12 + 1= 1

Check 1 is correct.

Verification 2

5x + 3y + 4z= 2 (2)

5(-4) + 3(6) + 4(1)=

-20 + 18 + 4= 2

Check 2 is correct.

Check 3

x + y - z= 1 (3)

1(-4) + 1(6) - 1(1)=

-4 + 6 - 1= 1

Check 3 is correct.

Exercise 2

5x - 3y - z= 1 (1)

x + 4y - 6z= -1 (2)

2x + 3y + 4z= 9 (3) **Solution:**

We solve this system of equations with 3 unknowns by the method of addition and subtraction as follows:

The first equation will be taken and the third one will eliminate the unknown Y

5x - 3y - z= 1 (1)

2x + 3y + 4z= 9 (3)

7x + 3z= 10 (4)

Equation 2 and 3 of the system will be taken and the unknown Y

x + 4y - 6z= -1 (2) 2x

+ 3y +4z= 9 (3)

Equation two will be multiplied by 3 and equation three will be multiplied by -4

$3x + 12y - 18z = -3$

$-8x - 12y - 16z = -36$

$-5x - 34z = -39 \ (5)$

We work with equation 4 and 5

$7x + 3z = 10 \ (4)$

$-5x - 34z = -39 \ (5)$

Equation four will be multiplied by -5 and the fifth equation by -7

$-35x - 15z = -50$

$35x + 238z = 273$

$223z =$

$$\frac{223z}{223} = \frac{223}{223}$$

$z = 1$

The value of z will be substituted in the fourth equation

$7x + 3z = 10 \ (4)$

$7x + 3(1) = 10$

$7x + 3 = 10$

$7x = 10 - 3$

$7x = 7$

$$\frac{7x}{7} = \frac{7}{7}$$

$x = 1$

Substitute the value of x,z in equation 3 of the system.

$2x + 3y + 4z = 9 \ (3)$

$2(1) + 3y + 4(1) = 9$

$2 + 3y + 4 = 9$

$3y = 9 - 2 - 4$

$3y =$

$$\frac{3y}{3} = \frac{3}{3}$$

$y = 1$

Answer: y=1, z=1 Verification

$5x - 3y - z = 1 \ (1)$

$5(1) - 3(1) - 1(1) =$

$5 - 3 - 1 = 1$

Check 1 is correct.

Verification 2

$x + 4y - 6z = -1 \ (2)$

$1(1) + 4(1) - 6(1) =$

$1 + 4 - 6 = -1$

Check 2 is correct.
Verification 3
2x + 3y + 4z= 9 (3)
2(1) + 3(1) + 4(1)=
2 + 3 + 4 + 4= 9
Check 3 is correct.

Exercise 3
2x - y + 2z= 6 (1)
3x + 2y - z= 4 (2)
4x + 3y - 3z= 1 (3)
Solution:
We solve this system of equations with 3 unknowns by the method of addition and subtraction as follows:
Equation one and three will be taken in which the unknown Y is eliminated
2x - y + 2z= 6 (1)
4x + 3y - 3z= 1 (3)
Equation 1 is multiplied by 3
6x - 3y + 6z= 18
4x + 3y - 3z= 1
10x + 3z= 19 (4)
Equation two and three are taken, the unknown Y
3x + 2y - z= 4 (2)
4x + 3y -3z= 1 (3)
Equation 2 is multiplied by 3 and the third equation by -2
9x + 6y - 3z= 12
-8x - 6y + 6z= -2
x + 3z= 10 (5)
We work with the equation 4 and 5
10x + 3z= 19 (4)
x + 3z= 10 (5)
Equation 5 is multiplied by -1
10x + 3z= 19
-x - 3z= -10
9x=
$$\frac{9x}{9} = \frac{9}{9}$$
x= 1
Substitute the value of x in equation four
10x + 3z= 19 (4)
10(1) + 3z= 19
10 + 3z= 19

3z= 19 - 10

3z=

$$\frac{3z}{9\ 3} = \frac{9}{3}$$

z= 3

Substitute the value of x, z in equation 1 of the system.

2x - y + 2z= 6 (1)

2(1) - y + 2(3)= 6

2 - y + 6= 6

-y= 6 - 6 - 6 - 2

-y=

$$\frac{-y}{-2\ -1} = \frac{-2}{-1}$$

y= 2

Answer: x= 1, y= 2, z= 3

Check 1

2x - y + 2z= 6 (1)

2(

1

) - 1(2) + 2

(3)=

2 - 2 + 6= 6

Check 1 is correct.

Verification 2

3x + 2y - z= 4 (2)

3(1) + 2

(2) - 1(3)=

3 + 4 - 3= 4

Check 2 is correct.

Verification 3

4x + 3y - 3z= 1 (3)

4(1) + 3(2) - 3(3)=

4 + 6 - 9= 1

Check 3 is correct.

Exercise 4

x + 2y + z= 0 (1)

2x + y + z= 1 (2)

x + y - z= 2 (3)

Solution:

We solve this system of equations with 3 unknowns by the method of addition and subtraction as follows:

Equation 1,3 is taken and the unknown z is eliminated
$x + 2y + z = 0$ (1)
$x + y - z = 2$ (3)
$2x + 3y = 2$ (4)
Equation two and three will have the unknown z eliminated
$2x + y + z = 1$ (2)
$x + y - z = 2$ (3)
$3x + 2y = 3$ (5)
We work with equation four and five
$2x + 3y = 2$ (4)
$3x + 2y = 3$ (5)
Equation four is multiplied by 3 and equation five is multiplied by -2 $6x + 9y = 6$
$-6x - 4y = -6$
$5y =$
$$\frac{5y}{5} = \frac{0}{5}$$
$y = 0$
Equation five substitutes the value of Y in equation five
$3x + 2y = 3$ (5)
$3x + 2(0) = 3$
$3x + 0 = 3$
$3x = 3 -$
$$\frac{3x}{3} = \frac{3}{3}$$
$x = 1$
Substitute the value of X, Y in the equation three of the system
$x + y - z = 2$ (3)
$1(1) + 1(0) - z = 2$
$1 + 0 - z = 2$
$-z = 2 - 1 - 0$
$-z =$
$$\frac{-z}{-1} = \frac{1}{-1}$$
$z = -1$
Answer: X= 1, Y= 0, Z= -1
Check 1
$x + 2y + z = 0$ (1)
$1(1) 1(1) + 2(0) + 1(-1) =$
$1 + 0 - 1 = 0$
Check 1 is correct.
Verification 2
$2x + y + z = 1$ (2)

2(1) + 1(0) + 1(-1)=
2 + 0 - 1= 1
Check 2 is correct.
Verification 3
x + y - z= 2 (3)
1(1) + 1(0) - 1(-1)=
1 + 0 + 1= 2
Check 3 is correct.

20.2 Problems of system of equations with 3 unknowns by the method of addition and subtraction.

Problem 1:
A transport company manages a fleet of 60 trucks of 3 different models the biggest ones transport a daily average of 15000 kg and travel daily an average of 400 km, the medium ones transport an average of 10000 kg and travel 300 km and the small ones transport 5000 kg and travel 100 km on average daily so the company's trucks transport a total of 475 tons and travel 12500 km between them all. How many trucks does the company manage of each model?
Solution:
We solve this problem as follows:
Large trucks = x
Medium trucks = y
Small trucks = z We
then pose the system of equations with 3 unknowns.

x	+	y	+	z=	60	(1)
15000x	+	10000y	+	5000z=	475000	(2)
400x	+	300y	+	100z=	12500	

Equation two is divided by 1000 and equation three will be divided by 100

x	+	y	+	z=	60	(1)
15x	+	10y	+	5z=	475	(2)
4x	+	3y	+	z=	125	(3)

Then solve this system of equations with 3 unknowns by the method of addition and subtraction as follows:

Equation one and three are taken and the unknown Z is eliminated

x	+	y	+	z=	60	(1)
4x	+	3y	+	z=	125	(3)

Equation 1 is multiplied by -1

-x	-	y	-	z=	-60
4x	+	3y	+	z=	125
3x	+	2y	=	65	(4)

Equation two and three are taken, and the unknown Z is eliminated from them

15x	+	10y	+	5z=	475	(2)
4x	+	3y	+	z=	125	(3)

Equation three is multiplied by -5

15x	+	10y	+	5z= 475
-20x-	15y	-	5z=	-625
-5x	-	5y=	-150	(5)

We work with the equation four and five

3x	+	2y=	65	(4)
-5x	-	5y=	-150	(5)

Equation four is multiplied by -5 and equation five will be multiplied by -3

-15x	-	10y=	-325
15x	+	15y=	450
5y=			

$$\frac{5y}{5} = \frac{125}{5}$$

125

y= 25

We substitute the value of Y in equation four

3x	+	2y=	65	(4)
3x	+	2(25)=	65
3x	+	50=	65
3x	=	65 -	50
3x=			

$$\frac{3x}{3} = \frac{15}{3}$$

15

x= 5

Substitute the value of x,y in equation one of the system

x	+	y	+	z=	60	(1)
1(5)	+	1 (25)	+	z	=	60
5	+	25	+	z=	60
z=	60	z=	60 -	5 -	25
z=					30
x=					5
y=					25

Answer: There are 5 large trucks, 25 medium trucks and 30 small trucks.

Chapter 21: The laws of exponents.

The law°1 of the laws of power exponents raised to 0

Any number raised to the power 0 will always result in 1.

Proof that any number raised to the power 0 will always be 1:

$$\frac{2^2}{2^2} = \quad 2^0 = 1$$

Examples of the law °1 of the laws of exponents.

Example 1:
$5^0 = 1$

Example 2:
$100^0 = 1$

Example 3
: $\frac{1}{2}^0 = 1$

Example 4:
$3^0 = 1$

Example 5:
$10^0 = 1$

21.2 Law °2 of the laws of exponents power raised to 1

When raising any base to the exponent 1, the result will give the same base.

Example 1:
$5^1 = 5$

Example 2:
$100^1 = 100$

Example 3:
$2^1 = 2$

Example 4:
$20^1 = 20$

Example 5:
$6^1 = 6$

21.3 Law °3 of the laws of exponents (multiplication of exponents with the same base)

In the multiplication of exponents with the same base what happens is that the exponents will be added.

Examples of multiplication of exponents with the same base.
Example 1:
$2^2 \times 2^1 = 2^3$
We solve this multiplication of exponents with the same base as follows:
$2^2 = 2 \times 2 = 4$
$2^1 = 2$
$4 \times 2 = 8$
Answer: 2^3

Example 2:
$5^2 \times 5^3 = 5^5$
It is solved as follows:
$5^2 = 5 \times 5 = 25$
$5^3 = 5 \times 5 \times 5 \times 5 = 125$
$25 \times 125 = 3125$
Answer: 5^5

Example 3:
$2^3 \times 2^2 = 2^5$
Solved as follows:
$2^3 = 2 \times 2 \times 2 \times 2 = 8$
$2^2 = 2 \times 2 = 4$
$8 \times 4 = 32$
Answer: 2^5

Example 4:
$3^3 \times 3^2 = 3^5$
It is solved as follows:
$3^3 = 3 \times 3 \times 3 \times 3 = 27$
$3^2 = 3 \times 3 = 9$
$27 \times 9 = 243$
Answer: 3^5

Example 5:
$4^3 \times 4^4 = 4^7$
It is solved as follows:

$4^3 = 4 \times 4 \times 4 = 64$
$4^4 = 4 \times 4 \times 4 \times 4 \times 4 \times 4 = 256$
$64 \times 256 = 16834$
Answer: 4^7

21.4 Law °4 of the laws of exponents (division of exponents with the same base)

In the division of exponents with the same base the exponents are subtracted.

$$\frac{a^m}{a^n} = a^{m-n}$$

Examples of division of exponents with the same base.

Example 1: $\dfrac{5^5}{5^2} = 5^3$

To be solved as follows:
$5^5 = 5 \times 5 \times 5 \times 5 \times 5 \times 5 = 3125$
$5^2 = 5 \times 5 \times 5 =$
$\dfrac{3125}{25} = 125$
25   25

Answer: 5^3

Example 2:
$\dfrac{2^6}{2^4} = 2^2$
To be solved as follows:

$2^6 = 2 \times 2 \times 2 \times 2 \times 2 \times 2 \times 2 \times 2 \times 2 \times 2= 64$

$2^4 = 2 \times 2 \times 2 \times 2 \times 2 \times 2=$

$$\frac{64}{16} \quad \frac{}{16} = 4$$

Answer: 2^2

Example 3

$$\frac{3^5}{3^2} = 3^3$$

To be solved as follows:

$3^5 = 3 \times 3 \times 3 \times 3 \times 3 \times 3 \times 3= 243$

$3^2 = 3 \times 3 \times 3=$

$$\frac{243}{9} \quad \frac{}{9} = 27$$

Answer: 3^3

Example 4

$$\frac{x^{10}}{x^6} = \quad x^4$$

How to solve this division of exponents with the same base is as follows.

$x^{10} = x \times x$

Then do the following.

$x^6 = x \times x \times x \times x \times x \times x \times x \times x \times x \times x \times x \times x$

Answer: x^4

Example 5

$$\frac{y^5}{y^2} = \quad y^3$$

To be solved as follows:

$y^5 = y \times y \times y \times y \times y \times y \times y$

$and^2 = y \times y\, y$

Answer: y^3

21.5 Law °5 of the laws of exponents(negative power)

Examples of a negative power.
Example 1:
2^{-2}
To be solved as follows:
$$\frac{1}{2^2} = \frac{1}{4}$$

Answer: $\frac{1}{4}$

Example 2:
5^{-3}
It will be solved as follows:
$$\frac{1}{5^3} = \frac{1}{125}$$

Response: $\dfrac{1}{125}$

Example 3:
2 $^{-5}$
It will be solved as follows:
$$\dfrac{1}{2^5} = \dfrac{1}{32}$$

Answer: $\dfrac{1}{32}$

Example 4:
10 $^{-4}$
It will be solved as follows:
$$\dfrac{1}{10^4} = \dfrac{1}{10000}$$

Answer: $\dfrac{1}{10000}$

Example 5:
4 $^{-3}$
It will be solved as follows: $\dfrac{1}{4^3} = \dfrac{1}{64}$

Response: $\dfrac{1}{64}$

21.6 Law °6 of the laws of exponents (power raised to another power)

In this law what happens is that the result will be the same base and the exponents will   be   multiplied.

Examples   of   power   raised   to   another   power.
Example   1:
(2   ) ^{23}
It   will   be   solved   as   follows:
2 ^{3} $^×$ ^{2}
2 ^{6}

Answer: 2^6

Example 2:
$(x^5)^2$
To be solved as follows:
$x^{5 \times 2}$
x^{10}
Answer: x^{10}

Example 3:
$(3^2)^4$
To be solved as follows:
$3^{2 \times 4}$
3^8
Answer: 3^8

Example 4:
$(y^2)^7$
To be solved as follows:
$y^{2 \times 7}$
y^{14}
Answer: y^{14}

Example 5:
$(4^3)^4$
To be solved as follows:
$4^{3 \times 4}$
4^{12}
Answer: 4^{12}

21.7 Law °7 of the laws of exponents(product raised to a power)

When a product of two or more factors is all raised to a power, the result will be the same product with each factor raised to the given exponent.

Examples of a product raised to a power.
Example 1:
$(xy)^2$
To be solved as follows:

x^2

Answer: $\qquad$ x^2 $\qquad$ y $\qquad$ y^2
$\qquad$ 2

PA

Example $\qquad$ **2:**
(bcd) ^{3}

To $\qquad$ be $\qquad$ solved $\qquad$ as $\qquad$ follows:

b^3 $\qquad$ c^3 $\qquad$ d^3

Answer: $\qquad$ b^3 $\qquad$ c^3 $\qquad$ d $\qquad$ 3

Example $\qquad$ **3:**
$(2x)$ ^{3}

To $\qquad$ be $\qquad$ solved $\qquad$ as $\qquad$ follows:

2^3 $\qquad$ x^3

8 $\qquad$ x^3

Answer: $\qquad$ 8 $\qquad$ x $\qquad$ 3

Example $\qquad$ **4:**
$(5yz)$ ^{3}

To $\qquad$ be $\qquad$ solved $\qquad$ as $\qquad$ follows:

5^3 $\qquad$ y^3 $\qquad$ z^3

125 $\qquad$ y^3 $\qquad$ z^3

Answer: $\qquad$ 125 $\qquad$ y^3 $\qquad$ z $\qquad$ 3

Example $\qquad$ **5:**
(abc) ^{4}

To $\qquad$ be $\qquad$ solved $\qquad$ as $\qquad$ follows:

a^4 $\qquad$ b^4 $\qquad$ c^4

Answer: $\qquad$ a^4 $\qquad$ b^4 $\qquad$ c $\qquad$ 4

21.8 Law °8 of the laws of exponents (Quotient raised to a power)

When a quotient is raised all at once to a power, the result will be the same quotient but the dividend and divisor will be raised to the given power.

Example $\qquad$ **1:**

$$\left(\frac{(x)^2}{(3)^2} \right)$$

It will be resolved as follows:

$$\frac{x^2}{3^2} = \frac{x^2}{9}$$

PA

Answer: $\frac{x^2}{9}$

Example 2: $\left(\frac{y^3}{2}\right)_4$

It will be solved as follows: $\frac{y^{12}}{2^4} = \frac{y^{12}}{16}$

Answer: $\frac{y^{12}}{16}$

Example 3: $\frac{x^2}{2})_3$

$($

It will be solved as follows: $\frac{x^6}{2^3} = \frac{x^6}{8}$

Answer: $\frac{x^6}{8}$

Example 4: $\frac{z^5}{5})_2$

$($

It will be solved as follows:

$$\frac{z^{10}}{5^2} = \frac{z^{10}}{25}$$

Answer: $\frac{z^{10}}{25}$

Chapter 22: Laws of radicals.
Law °1 of radicals and law of cancellation.

Cancellation law.

$$(\sqrt[n]{a^n} = a$$

Examples of the law of cancellation
Example 1:

$$\sqrt[2]{2^2} = 2$$

It will be solved as follows:
The square of the square root cancels with the square found in the radical.
Answer: 2

Example 2

$$: \sqrt[2]{5^2} = 5$$

To be solved as follows:
The square of the square root cancels with the square found in the radical.
Answer: 5

Example 3: $\sqrt[2]{10^2} = 10$

To be solved as follows:
The square of the square root cancels with the square found in the radical.
Answer: 10

Example 4

$$: \sqrt[3]{4^3} = 4$$

It will be solved as follows:
The square of the square root cancels with the square found in the radical.
Answer: 4

Example 5: $\sqrt[3]{20^3} = 20$

It will be solved as follows:
The square of the square root cancels with the square found in the radical.
Answer:20

22.2 Law °2 of the laws of radicals (root of a product).

Root law of a product.

$$\sqrt[n]{ab} = \sqrt[n]{a}\ \sqrt[n]{b}$$

Examples of the root law of a product.

Example 1

$\sqrt[2]{18} = 3\sqrt{2}$

It is solved as follows: $\sqrt[2]{9 \times 2}$

Then we take out the 9 that is inside the radical, since we can take the exact square root of the 9.

$3\sqrt{2}$

Answer: $3\sqrt{2}$

Example 2:

$\sqrt[2]{8} = 2\sqrt{2}$

It is solved as follows:

$\sqrt[2]{4 \times 2}$

Then the 4 that is inside the radical is taken out, since exact square root can be taken from 4.

$2\sqrt{2}$

Answer: $2\sqrt{2}$

Example 3: $\sqrt[2]{32} = 4\sqrt{2}$

It is resolved as follows: $\sqrt[2]{16 \times 2}$

Then we take out the 16 that is inside the radical, since we can take the exact square root.

$4\sqrt{2}$

Answer: $4\sqrt{2}$

Example 4: $\sqrt[2]{50} = 5\sqrt{2}$

It will be resolved as follows:

$\sqrt[2]{25 \times 2}$

Then we take out the 25 that is inside the radical, since we can take the exact square root.

$5\sqrt{2}$

Answer: $5\sqrt{2}$

Example 5: $\sqrt[2]{72} = 6\sqrt{2}$

It will be resolved as follows: $\sqrt[2]{36 \times 2}$

Then we take out the 36 that is inside the radical, since we can take the exact square root.

$6\sqrt{2}$

Answer: $6\sqrt{2}$

22.3 Law °3 of the laws of radicals (root of a division)

Root law of a division.

$$\sqrt[n]{\frac{a}{b}} = \frac{\sqrt[n]{a}}{\sqrt[n]{b}}$$

Examples of the law

Example 1

$$\sqrt[2]{\frac{9}{4}} = \frac{3}{2}$$

It is resolved as follows:

$$\frac{\sqrt[2]{9}}{\sqrt[2]{4}} = \frac{3}{2}$$

Response: $\dfrac{3}{2}$

Example 2

$: \sqrt[2]{\dfrac{36}{25}} = \dfrac{6}{5}$

It is solved as follows:

$\dfrac{\sqrt[2]{36}}{\sqrt[2]{25}} = \dfrac{6}{5}$

Answer: $\dfrac{6}{5}$

Example 3: $\sqrt[2]{\dfrac{100}{81}} = \dfrac{10}{9}$

It is resolved as follows: $\dfrac{\sqrt[2]{100}}{\sqrt[2]{81}} = \dfrac{10}{9}$

Response: $\dfrac{10}{9}$

Example 4

$: \sqrt[3]{\dfrac{125}{27}} = \dfrac{5}{3}$

It is resolved as follows:

$\dfrac{\sqrt[3]{125}}{\sqrt[3]{27}} = \dfrac{5}{3}$

Answer: $\dfrac{5}{3}$

Example 5

$: \sqrt[3]{\dfrac{1000}{8}} = 5$

It is resolved as follows:

$\dfrac{\sqrt[3]{1000}}{\sqrt[3]{8}} = \dfrac{10}{2}$

$$\frac{10}{2} = 5$$

Answer: 5

22.4 Law °4 of the laws of radicals (root of a root)

Root law of a root.

$$\sqrt[m]{\sqrt[n]{a}} = \sqrt[mn]{a}$$

Examples of the law

Example 1: $\sqrt[2]{\sqrt[6]{21}} = \sqrt[12]{21}$

It is solved as follows:

The indices of the roots are multiplied

2 × 6= 12

Radical=

21. $\sqrt[12]{21}$

Answer: $\sqrt[12]{21}$

Example 2

: $\sqrt[2]{\sqrt[3]{12}}$ = $\sqrt[6]{12}$

It is solved as follows:

The indices of the roots are multiplied

2 × 3= 6

Radical=

12. $\sqrt[6]{12}$

   Answer: $\sqrt[6]{12}$

Example 3: $\sqrt[2]{\sqrt[2]{16}}$ **= 2**

It is solved as follows:

The indices of the roots are multiplied

2 × 2= 4

Radical=16 $\sqrt[4]{16}$ **= 2**

Answer: 2

Example 4: $\sqrt[2]{\sqrt[2]{10}}$ = $\sqrt[4]{10}$

It is solved as follows:

The indices of the roots are multiplied

2 × 2= 4

Radical=

10. $\sqrt[4]{10}$

 Answer: $\sqrt[4]{10}$

Example 5: $\sqrt[3]{\sqrt[2]{2}}$ = $\sqrt[6]{2}$

It is solved as follows:

The indices of the roots multiply

3 × 2= 6

Radical=

2 $\sqrt[6]{2}$

   Answer: $\sqrt[6]{2}$

22.5 Law °5 of the laws of radicals(root of a power)

Root law of a power

$\sqrt[n]{a^m} = a^{\frac{m}{n}}$

Examples of the root law of a power.
Example 1

$\sqrt[3]{5^6} = 25$

It is solved as follows:
the following

division is made $\frac{6}{3} = 2$

$5^2 = 25$
Answer:25

Example 2

$: \sqrt[5]{10^{10}} = 100$

It is solved in the following way:

Divide $\dfrac{10}{5} = 2$

$10^2 = 100$
Answer: 100

Example 3: $\sqrt[3]{2^9} = 8$
It is solved as follows:
The following

division is made $\dfrac{9}{3} = 3$

$2^3 = 8$
Answer:8

Example 4: $\sqrt[9]{3^{27}} = 27$

It is solved as follows:
the following

division is made $\dfrac{27}{9} = 3$

$3^3 = 27$
Answer: 27

Example 5

$: \sqrt[10]{2^{50}} = 32$

It is solved as follows:
the following

division is made $\dfrac{50}{10} = 5$

$2^5 = 32$
Answer: 32

Chapter 23: Binomial squared.
Demonstration of the formula for a binomial squared.

Demonstration of the formula for a positive squared binomial:
$(a + b)^2$
$(a + b)(a + b)$
This multiplication is solved as follows:
The first term in the first parenthesis multiplies all the terms in the second parenthesis.
a × a= a^2
a × b= ab.
The second term of the first parenthesis multiplies all the terms of the second parenthesis.

b × a= ab
b × b= b²
 a² + ab + ab + b²
 a² + 2ab + b ²
This formula will be used to solve a positive squared binomial.
The formula is applied as follows:
The first term of the positive squared binomial plus twice the first term times the second term plus the last term squared.

Demonstration of the formula for a negative squared binomial.
(a - b)²
 (a - b)(a - b)
This multiplication is solved as follows:
The first term of the first parenthesis multiplies all the terms of the second parenthesis.
a × a= a²
 a × -b= -ab
The second term of the first parenthesis is multiplied by all the terms of the second parenthesis.
-b × a=-ab
-b × -b= b ²
a² - ab - ab + b²
 a² - 2ab + b ²
This is the formula for a negative squared binomial.

23.2 Examples of the binomial squared.

Examples of the positive squared binomial.
Example 1:
(x + y) ²
It is solved in the following way:
(x + y
)(x + y
)
The first term in the first parenthesis is multiplied by all the terms in the second parenthesis.
x × x= x²
 x × y= xy

The second term in the first parenthesis is multiplied by all the terms in the second parenthesis.

$y \times x = xy$

$y \times y = y^2$

$x^2 + xy + xy + y^2$

$x^2 + 2xy + y^2$

Answer: $x^2 + 2xy + y^2$

Example 2:

$(x + 3)^2$

It is solved as follows:

$(x + 3$

$)(x + 3)$

The first term in the first parenthesis is multiplied by all the terms in the second parenthesis.

$x \times x = x^2$

$x \times 3 = 3x.$

The second term in the first parenthesis is multiplied by all the terms in the second parenthesis.

$3 \times x \times x = 3x$

$3 \times 3 = 9$

$x^2 + 3x + 3x + 3x + 9$

$x^2 + 6x + 9$

Answer: $x^2 + 6x + 9$

Example 3:

$(y + 2)^2$

It is solved as follows: (

$y + 2$

$)(y + 2$

$)$

The first term in the first parenthesis is multiplied by all the terms in the second parenthesis.

$y \times y = y^2$

$y \times 2 = 2y$

The second term in the first parenthesis is multiplied by all the terms in the second parenthesis.

$2 \times y = 2y$

2 × 2= 4
$y^2 + 2y + 2y + 2y + 4$
$y^2 + 4y + 4$
Answer: $y^2 + 4y + 4$

Example 4:
$(z + 5)^2$
It will be solved as follows: (
z + 5
)(z + 5
)
The first term in the first parenthesis is multiplied by all the terms in the second parenthesis.
$z \times z = z^2$
$z \times 5 = 5z$
The second term in the first parenthesis is multiplied by all the terms in the second parenthesis.
$5 \times z = 5z$
$5 \times 5 = 25$
$z^2 + 5z + 5z + 25$
$z^2 + 10z + 25$
Answer: $z^2 + 10z + 25$

Example 5:
$(w + 10)^2$
It will be solved as follows: (
w + 10)(w + 10)
The first term of the first parenthesis is multiplied by all the terms of the second parenthesis.
$w \times w = w^2$
$w \times 10 = 10w$
The second term in the first parenthesis is multiplied by all the terms in the second parenthesis.
$10 \times w = 10w$
$10 \times 10 = 100$
$w^2 + 10w + 10w + 100$
$w^2 + 20w + 100$
Answer: $w^2 + 20w + 100$

Examples of the negative squared binomial.
Example 1:
$(x - 5)^2$
It is solved in the following way:

(x - 5
)(x - 5
)
The first term of the first parenthesis is multiplied by all the terms of the second parenthesis.

$x \times x = x^2$

$x \times -5 = -5x$

The second term in the first parenthesis is multiplied by all the terms in the second parenthesis.

$-5 \times x = -5x$

$-5x \ -5 \times -5 = 25$

$x^2 - 5x - 5x + 25$

$x^2 - 10x + 25$

Answer: $x^2 - 10x + 25$

Example 2:

$(2x - 3)^2$

It is solved as follows:

(2x - 3
)(2x - 3
)
The first term in the first parenthesis is multiplied by all the terms in the second parenthesis.

$2x \times 2x = 4x^2$

$2x \times 2x \times -3 = -6x$

The second term in the first parenthesis is multiplied by all the terms in the second parenthesis.

$-3 \times 2x = -6x$

$-3 \times -3 = 9$

$4x^2 - 6x - 6x + 9$

$4x^2 - 12x + 9$

Answer: $4x^2 - 12x + 9$

Example 3:

$(5x - 10)^2$

To be solved as follows:

(5x - 10)(5x - 10)

The first term in the first parenthesis is multiplied by all the terms in the second parenthesis.

$5x \times 5x = 25x^2$

$5x \times -10 = -50x$

The second term in the first parenthesis is multiplied by all the terms in the second parenthesis.

-10 × 5x= -50x.
-10 × -10= 100
$25x^2 - 50x - 50x + 100$
$25x^2 - 100x + 100$
Answer: $25x^2 - 100x + 100$

Example 4:
$(y - 4)^2$
To be solved as follows:
(y - 4
)(y - 4
)
The first term in the first parenthesis is multiplied by all the terms in the second parenthesis.
$y × y= y^2$
$y × -4= -4y$
The second term in the first parenthesis is multiplied by all the terms in the second parenthesis.
-4 × y= -4y
-4 × -4= 16
$y^2 - 4y - 4y + 16$
$y^2 - 8y + 16$
Answer: $y^2 - 8y + 16$

Example 5:
$(3z - 7)^2$
To be solved as follows:
(3z - 7
)(3z - 7
)
The first term in the first parenthesis is multiplied by all the terms in the second parenthesis.
$3z × 3z= 9z^2$
3z × -7= -21z
The second term of the first parenthesis is multiplied by all the terms of the second parenthesis.
-7 × 3z= -21z
-7 × -7= 49
$9z^2 - 21z - 21z + 49$
$9z^2 - 42z + 49$
Answer: $9z^2 - 42z + 49$

23.3 Binomial squared exercises.

Exercise 1
$(x + 1)^2 = x^2 + 2x + 1$

Exercise 2
$(2x + 2)^2 = 4x^2 + 8x + 4$

Exercise 3
$(xy + 2)^2 = x y^{22} + 4xy + 4$

Exercise 4
$(3z + 6)^2 = 9z^2 + 36z + 36$

Exercise 5
$(4x + 5)^2 = 16x^2 + 40x + 25$

Exercise 6
$(x - 8)^2 = x^2 - 16x + 64$

Exercise 7
$(y - 2)^2 = y^2 - 4y + 4$

Exercise 8
$(xy - 4)^2 = x y^{22} - 8xy + 16$

Exercise 9
$(z - 3)^2 = z^2 - 6z + 9$

Exercise 10
$(10w - 1)^2 \ 100w^2 - 20w + 1$

$(a + b)^3$

$(a + b)^2 (a + b)$

$(a + b)^2$

$(a + b)(a + b)$

This multiplication is solved in the following way:

We multiply the first term of the first parenthesis by all the terms of the second parenthesis.

$a \times a = a^2$

$a \times b = ab$

The second term of the first parenthesis will be multiplied by all the terms of the second parenthesis.

$b \times b = b^2$

$a^2 + ab + ab + b^2$

$a^2 + 2ab + b^2$

$(a^2 + 2ab + b^2)(a + b)$

The first term of the first parenthesis is multiplied by all the terms of the second parenthesis.

$a^2 \times a = a^3$

$a^2 \times b = a b^2$

The second term in the first parenthesis is multiplied by all the terms in the second parenthesis.

$2ab \times a = 2 a^2$

$b\ 2ab \times b = 2\ ab^2$

The third term of the first parenthesis is multiplied by all the terms of the second term

$b^2 \times a = ab^2$

$b^2 \times b = b^3$

$a^3 + a^2 b + 2 a^2 b + 2ab^2 + ab^2 + b^3$

$a^3 + 3 a^2 b + 3ab^2 + b^3$

The formula is applied as follows:

The first term of the binomial is cubed plus three times the first term squared times the second term plus three times the first term times the second term squared plus the last term cubed.

54.2 Examples of cubed binomial.

Example 1:
$(x + 1)^3$
To be solved as follows:
$x^3 + 3(x^2)(1) + 3(x)(1^2) + 1^3$
$x^3 + 3x^2 + 3x + 3x + 1$
Answer: $x^3 + 3x^2 + 3x + 1$

Example 2:
$(x + 3)^2$
To be solved as follows:
$x^3 + 3(x^2)(3) + 3(x)(3) + 3^3$
$x^3 + 9x^2 + 9x + 27$
Answer: $x^3 + 9x^2 + 9x + 27$

Example 3:
$(y + 5)^3$
To be solved as follows:
$y^3 + 3(y^2)(5) + 3(y)(5^2) + 5^3$
$y^3 + 15y^2 + 75y + 125$
Answer: $y^3 + 15y^2 + 75y + 125$

Example 4:
$(z + 2)^3$
To be solved as follows:
$z^3 + 3(z^2)(2) + 3(z)(2^2) + 2^3$
$z^3 + 6z^2 + 12z + 8$
Answer: $z^3 + 6z^2 + 12z + 8$

Example 5:
$(w + 10)^3$
To be solved as follows:
$w^3 + 3(w^2)(10) + 3(w)(10^2) + 10^3$
$w^3 + 30w^2 + 300w + 1000$
Answer: $w^3 + 30w^2 + 300w + 1000$

54.3 Binomial cubed exercises.

PA

Exercise 1
$(x + 4)^3 = x^3 + 12x^2 + 48x + 64$

Exercise 2
$(y + 6)^3 = y^3 + 18y^2 + 108y + 216$

Exercise 3
$(2x + 1)^3 = 8x^3 + 12x^2 + 6x + 1$

Exercise 4
$(5y + 2)^3 = 125y^3 + 150y^2 + 60y + 8$

Exercise 5
$(3z + 5)^3 = 27z^3 + 45z^2 + 225z + 125$

Examples of conversion from degrees to radians.

To convert from degrees to radians you have to know that one complete revolution of a circle is 360 degrees. Knowing that a half turn of a circle is 180 degrees, this will be taken as a reference to be able to convert from degrees to radians.

Example 1: Convert 60 degrees to radians.

It will be solved as follows:

$$\frac{60 \text{ grados}}{1} \times \frac{\text{pi radianes}}{180 \text{ grados}} = \frac{60\text{pi}}{180} \text{ radians}$$

Then do the following

$$\frac{60\text{pi}}{180} \text{ radians} = \frac{30\text{pi}}{90} \text{ radians}$$

$$\frac{30\text{pi}}{90} \text{ radians} = \frac{15\text{pi}}{45} \text{ radian s}$$

$$\frac{15\text{pi}}{45} \text{ radians} = \frac{3\text{pi}}{9} \text{ radian s}$$

$$\frac{3\text{pi}}{9} \text{ radians} = \frac{\text{pi}}{3} \text{radians}$$

Answer: 60 degrees = $\frac{\text{pi}}{3}$ radians.

Example 2:

Convert 30 degrees to radians.

It will be solved as follows:

$$\frac{30 \text{ grados}}{1} \times \frac{\text{pi radianes}}{180 \text{ grados}} = \frac{30\text{pi}}{180} \text{ radian s}$$

$$\frac{30\text{pi}}{180} \text{ radians} = \frac{15\text{pi}}{90} \text{ radian s}$$

$$\frac{15\text{pi}}{90} = \frac{3\text{pi}}{18} \text{ radian s}$$

$$\frac{3\text{pi}}{18} \text{radians} = \frac{\text{pi}}{6} \text{radians}$$

Answer: 30 degrees = $\frac{\text{pi}}{6}$ radians

Example 3:

Convert **45** **degrees** **to** **radians**

It will be solved as follows: $\dfrac{45 \text{ grados}}{1} \times \dfrac{\text{pi radianes}}{180 \text{ grados}} = \dfrac{45\text{pi}}{180}$ radian

s $\dfrac{45\text{pi}}{180}$ radians = $\dfrac{9\text{pi}}{36}$ radians

$\dfrac{9\text{pi}}{36}$ radians = $\dfrac{\text{pi}}{4}$ radians

Answer: 45 degrees = $\dfrac{\text{pi}}{4}$ **radians**

Example 4:
Converting **90** **degrees** **to** **radians**

It will be solved as follows: $\dfrac{90 \text{ grados}}{1} \times \dfrac{\text{pi radianes}}{180 \text{ grados}}$

$\dfrac{90 \text{ pi}}{180}$ radian

s $\dfrac{90\text{pi}}{180} = \dfrac{45\text{pi}}{90}$ radian

s $\dfrac{45\text{pi}}{90}$ radians

$\dfrac{9\text{pi}}{18}$ radian

s $\dfrac{9\text{pi}}{18}$ radians = $\dfrac{\text{pi}}{2}$ radians **Answer: 90 degrees =** $\dfrac{\text{pi}}{2}$ **radians**

Example 5:
Converting 240 degrees to radians
It will be solved as follows:
$\dfrac{240 \text{ grados}}{1} \times \dfrac{\text{pi radianes}}{180 \text{ grados}}$

$\dfrac{240\text{pi}}{180}$ radian

$\dfrac{120\text{pi}}{90}$ radianes

s

$$\frac{120pi}{90} \text{ radian}$$

s $$\frac{60 \text{ pi}}{45} \text{ radian}$$

s $$\frac{60 \text{ pi}}{45} \text{ radian}$$

s $$\frac{20pi}{15} \text{ radian}$$

s $$\frac{20pi}{15} \text{ radian}$$

s $$\frac{4pi}{3} \text{ radians}$$

Answer: 240 degrees = $\frac{4pi}{3}$ radians

25.2 Exercises on converting degrees to radians

Exercise 1
Convert 360 degrees to radians

R= 2pi radians

Exercise 2
Converting 180 degrees to radians

R= pi radians

Exercise 3
Converting 120 degrees to radians

R= $\frac{2pi}{3}$ radians

Exercise 4
Converting 270 degrees to radians

R= $\dfrac{3pi}{2}$ radians

Exercise 5 Convert 720 degrees to radians

R= 4pi radians

25.3 Examples of radians to degrees conversion

Example 1:

Converting $\dfrac{pi}{2}$ radians to degrees

It is solved as follows: $\dfrac{180 \ grados}{2} \times 1$

90 degrees × 1= 90 degrees

Answer: $\dfrac{pi}{2}$ radians = 90 degrees

Example 2:

Converting $\dfrac{pi}{4}$ radians to degrees

It is solved as follows: $\dfrac{180 \ grados}{4} \times 1$

45 degrees × 1= 45 degrees

Answer: $\dfrac{pi}{4}$ radians = 45 degrees

Example 3:

Converting $\dfrac{pi}{6}$ **radians to degrees**

It is solved as follows: $\dfrac{180 \text{ grados}}{6} \times 1$

30 degrees × 1= 30 degrees

Answer: $\dfrac{pi}{6}$ **radians = 30 degrees**

Example 4:

Converting $\dfrac{pi}{3}$ **radians to degrees**

It is solved as follows: $\dfrac{180 \text{ grados}}{3} \times 1$

60 degrees × 1= 60 degrees

Answer: $\dfrac{pi}{3}$ **radians = 60 degrees**

Example 5:

Converting $\dfrac{7pi}{3}$ **radians to degrees**

It is solved as follows:

$\dfrac{180 \text{ grados}}{3} \times 7$

60 degrees × 7= 420 degrees

Answer: $\dfrac{7pi}{3} = 420$ **degrees**

Exercises

Exercise 1 Convert $\dfrac{2pi}{3}$ **radians to degrees**

R= 120 degrees

Exercise 2

Convert $\dfrac{3pi}{2}$ radians to degrees

R= 270 degrees

Exercise 3

Convert $\dfrac{4pi}{3}$ radians to degrees

R= 240 degrees

Exercise 4

Convert $\dfrac{5pi}{2}$ radians to degrees

R= 450 degrees

Exercise 5

Convert $\dfrac{8pi}{3}$ radians to degrees

R= 480 degrees

Chapter 26: Converting centimeters to meters
Examples of conversion from centimeters to meters

Example 1:
Converting 50 cm to meters

It will be solved as follows:
Knowing that 1m= 100cm
A

$$\frac{50cm}{1} \times \frac{1m}{100cm}$$

conversion factor

As you have centimeters above and below, these will be

$$\frac{50}{100} = 0.5m$$

eliminated.

Answer: 50cm = 0.5m

Example 2
Convert 20 cm to meters
It will be solved as follows: **Knowing that 1m= 100cm**
We make a

$$\frac{20cm}{1} \times \frac{1m}{100cm}$$

conversion factor

$$\frac{20}{100} = 0.2 \text{ m}$$

Answer: 20cm= 0.2m

Example 3:
Converting 45 cm to meters
It will be solved as follows:
Knowing that 1m=

$$\frac{45cm}{1} \times \frac{1m}{100cm}$$

100cm

$$\frac{45}{100} = 0.45 \text{ m}$$ **Answer: 45cm= 0.45m**

Example 4:
Converting 25 cm to meters
It will be solved as follows:

$$\frac{25cm}{1} \times \frac{1m}{100cm}$$

$$\frac{25}{100} = 0.25 \text{ m}$$

**Answer: 25cm = 0
.25m**

26.2 Exercises on converting centimeters to meters

Exercise 1
Converting 10 cm to meters

R= 0.1m

Exercise 2
Convert 18 cm to meters

R= 0.18m

Exercise 3
Converting 65 cm to meters

R= 0.65m

Exercise 4
Converting 82 cm to meters

R= 0.82m

Exercise 5
Converting 96 cm to meters

R= 0.96m

PA

Example 1:
Converting 25,000 cm² to m²
This exercise is solved as follows:
By making a

conversion factor $\dfrac{25000cm^2}{1} \times (\dfrac{1m}{100})^2$

$\dfrac{25000cm^2}{1} \times \dfrac{1m^2}{10000cm^2}$

You have cm² above and cm² below, then the cm² are

eliminated $\dfrac{25000}{10000} = 2.5$ m²

Answer: 25,000 cm² = 2.5m²

Example 2:
Converting 15,000 cm² to m²
This exercise is solved as follows:
By making a

conversion factor $\dfrac{15000cm^2}{1} \times (\dfrac{1m}{100cm})^2$

$\dfrac{15000cm^2}{1} \times \dfrac{1m^2}{10000cm^2}$

$\dfrac{15000}{10000} = 1.5$ m²

Answer: 15,000 cm² = 1.5m²

Example 3:
Converting 30,000 cm² to m²
This exercise is solved as follows:
By making a

conversion factor $\dfrac{30000cm^2}{1} \times \left(\dfrac{1m}{100cm}\right)^2$

$$\frac{30000cm^2}{1} \times \frac{1m^2}{10000cm^2}$$

$$\frac{30000}{10000} = 3 \text{ m}^2$$

Answer: 30,000 cm² = 3m ²

Example 4:
Converting 45,000cm² to m ²
This exercise is solved as follows:
By making a

conversion factor $\frac{45000cm^2}{1} \times (\frac{1m}{100cm})^2$

$$\frac{45000cm^2}{1} \times \frac{1m^2}{10000cm^2}$$

$$\frac{45000}{10000} = 4.5m^2$$

Answer: 45,000 cm² = 4.5m ²

Example 5:
Convert 55,000 cm² to m ²
This exercise is solved as follows:
By making a

conversion factor $\frac{55000cm^2}{1} \times (\frac{1m}{100cm})^2$

$$\frac{55000cm^2}{1} \times \frac{1m^2}{10000cm^2}$$

$$\frac{55000}{10000} = 5.5m^2$$

Answer: 55,000 cm² = 5.5m ²

27.2 Exercises on converting square centimeters to square meters

Exercise 1
Convert 35,000 cm^2 to m^2

R= 3.5m^2

Exercise 2
Convert 60,000cm^2 to m^2

R= 6m^2

Exercise 3
Convert 72,000 cm^2 to m^2

R= 7.2m^2

Exercise 4
Convert 85,000 cm^2 to m^2

R= 8.5m^2

Exercise 5
Convert 94,000 cm^2 to m^2

R= 9.4m^2

27.3 Examples of cubic centimeters to cubic meters conversion

Example 1:
Converting 50,000 cm³ to m ³
This exercise is solved as follows:
By making a

conversion factor $\dfrac{50000cm^3}{1}$ $\times (\dfrac{1m}{100cm}$ ³$)$

$\dfrac{50000cm^3}{1}$ $\times \dfrac{1m^3}{1000000cm^3}$

You have cm³ above and cm³ below, then the cm³ will be

eliminated. $\dfrac{50000}{1000000} = $ 0.0 **5m³**

Answer: 50,000cm³ = 0.05m ³

Example 2:
Converting 25,000 cm³ to m ³
This exercise is solved as follows:
By making a

conversion factor $\dfrac{25000cm^3}{1}$ $\times \left(\dfrac{1m}{100cm}\right)$₃

$$\frac{25000cm^3}{1} \times \frac{1m^3}{1000000cm^3}$$

$$\frac{25000}{1000000} = 0.025 \ m^3$$

Answer: 25,000cm³ = 0.025m³

Example 3:
Converting 72,000cm³ to m³
This exercise is solved as follows:
By making a

conversion factor $\frac{72000cm^3}{1} \times \left(\frac{1m}{100cm}\right)^3$

$$\frac{72000cm^3}{1} \times \frac{1m^3}{1000000cm^3}$$

$$\frac{72000}{1000000} = 0.072 \ m^3$$

Answer: 72,000 cm³ = 0.072m³

Example 4:
Converting 3000000 cm³ to m³
This exercise is solved as follows:
By making a

conversion factor $\frac{3000000cm^3}{1} \times \frac{1}{(100cm)^3}$

$$\frac{3000000cm^3}{1} \times \frac{1m^3}{1000000cm^3}$$

$$\frac{3000000}{1000000} = 3m^3$$

Answer: 3000000cm³ = 3m³

Example 5:
Converting 10,000 cm³ to m³

This exercise is solved as follows:
By making a

conversion factor $\dfrac{\dfrac{10000cm^3}{1}}{} \times (\dfrac{1m}{100cm})^3$

$\dfrac{10000cm^3}{1} \times \dfrac{1m^3}{1000000cm^3}$

$\dfrac{10000}{1000000} = 0.01\ m^3$

Response: 10,000 cm³ = 0.01m ³

27.4 Cubic centimeter to cubic meter conversion exercises

Exercise 1
Convert 35,000 cm³ to m ³

R= 0.035m ³

Exercise 2
Convert 22,000 cm³ to m ³

R= 0.022m ^{3}

Exercise 3
Convert 67,000 cm³ to m ^{3}

R= 0.067m ^{3}

Exercise 4
Convert 72,000cm³ to m ^{3}

R= 7.2m ^{2}

Exercise 5
Convert 5000000 cm³ to m ^{3}

R= 5m ^{3}

Exercise 6
Convert 20,000 cm³ to m ^{3}

R= 0.02m ^{3}

Chapter 28: Conversion of kilometers to meters and conversion of meters to kilometers
Examples of conversion from kilometers to meters

Example 1:
Converting 12 km to meters

This exercise is solved as follows:
Knowing that 1 km= 1000m
A conversion factor will be made

$$\frac{12km}{1} \times \frac{1000m}{1km}$$

Since you have kilometers above and below these are eliminated
12 × 1000= 12000 m
Answer: 12 km= 12000 m
Example 2:
Converting 5.5 km to meters
This exercise is solved as follows:
Knowing that 1 km= 1000m

A conversion factor will be made $\dfrac{5.5km}{1} \times \dfrac{1000m}{1km}$

5.5 × 1000= 5500m
Answer: 5.5 km= 5500m

Example 3:
Converting 3.2 km to meters
This exercise is solved as follows:
Knowing that 1 km= 1000m

A conversion factor will be made $\dfrac{3.2km}{1} \times \dfrac{1000m}{1km}$

3.2 × 1000= 3200m
Answer: 3.2 km= 3200m

Example 4:
Converting 7.1 km to meters
This exercise is solved as follows:
Knowing that 1 km= 1000m
A conversion factor will be made

$$\frac{7.1km}{1} \times \frac{1000m}{1km}$$

7.1 × 1000= 7100 m
Answer: 7.1 km= 7100 m

Example 5:
Converting 8.5 km to meters
This exercise is solved as follows:

Knowing that 1 km= 1000m

A

conversion factor will be made $\dfrac{8.5km}{1}$ $\times$ $\dfrac{1000m}{1km}$

8.5 × 1000= 8500m

Answer: 8.5km= 8500m

28.2 Exercises on Converting Kilometers to Meters

Exercise 1
Converting 9.4 km to meters

R= 9400m

 Exercise 2
Convert 5.8 km to meters

R= 5800m

 Exercise 3
Convert 2.5 km to meters

R= 2500m

 Exercise 4
Convert 4 km to meters

R= 4000m

Exercise 5
Convert 1.6 km to meters

R= 1600m

28.3 Examples of meter to kilometer conversion

Example 1:

Convert 500 m to kilometers

This exercise is solved as follows:
Knowing that 1 km= 1000m

A conversion factor will be made $\dfrac{500m}{1} \times \dfrac{1km}{1000m}$

Since there are meters above and below, these will be
eliminated. $\dfrac{500}{1000} = 0.5$ km

Answer: 500 m= 0.5km

Example 2:
Converting 200 m to kilometers
This exercise is solved as follows:
Knowing that 1 km= 1000m
A

conversion factor will be made $\dfrac{200m}{1} \times \dfrac{1km}{1000m}$

$\dfrac{200}{1000} = 0.2$ km

Answer: 200 m= 0.2km

Example 3:
Converting 5000 m to kilometers
This exercise is solved as follows:
Knowing that 1 km= 1000m
A

conversion factor will be made $\dfrac{5000m}{1} \times \dfrac{1km}{1000m}$

$\dfrac{5000}{1000} = 5$ km

Answer: 5000 m= 5km

Example 4:
Converting 1500 m to kilometers
This exercise is solved as follows:
Knowing that 1 km= 1000m
A conversion factor will be made

$$\frac{1500m}{1} \times \frac{1km}{1000m}$$

$$\frac{1500}{1000} = 1.5 \ \textbf{km}$$

Answer: 1500m= 1.5km

Example 5:
Converting 7500 m to kilometers
This exercise is solved as follows:
Knowing that 1 km= 1000m
A

$$\frac{7500m}{1} \times \frac{1km}{1000m}$$

conversion factor will be made

$$\frac{7500}{1000} = 7.5 \ \textbf{km}$$

Answer: 7500 m= 7.5km

28.4 Exercises on converting meters to kilometers

Exercise 1
Converting 8200 m to kilometers

R= 8.2km

Exercise 2
Convert 4600 m to kilometers

R= 4.6km

Exercise 3
Converting 1300 m to kilometers

R= 1.3km

Exercise 4
Convert 9000 m to kilometers

R= 9km

Exercise 5
Convert 8200 m to kilometers

R= 8.2km

Chapter 29: Converting miles to kilometers and converting kilometers to miles
Examples of mile to kilometer conversions

Example 1: Converting 50 miles to kilometers

This exercise is solved as follows:
Knowing that 1 mi= 1.609 km
A

$$\frac{50mi}{1} \times \frac{1.609km}{1mi}$$

conversion factor will be made
50 × 1.609= 80.45 km
Answer: 50 mi= 80.45 km

Example 2:
Converting 100 miles to kilometers
This exercise is solved as follows:
Knowing that 1 mi= 1.609 km
A

$$\frac{100mi}{1} \times \frac{1.609km}{1mi}$$

conversion factor will be made
100 × 1.609= 160.9km
Answer: 100 mi= 160.9 km

Example 3:
Converting 25 miles to kilometers
This exercise is solved as follows:
Knowing that 1 mi= 1.609 km
A

$$\frac{25mi}{1} \times \frac{1.609km}{1}$$

conversion factor will be made
25 × 1.609= 40.225km
Answer: 25 mi= 40.225 km

Example 4:
Converting 10 miles to kilometers
This exercise is solved as follows:
 Knowing that 1 mi= 1.609 km
A

$$\frac{10mi}{1} \times \frac{1.609km}{1mi}$$

conversion factor will be made
 10 × 1.609= 6
.09km
Answer: 10 mi= 16.09 km

 Example 5:
Converting 1000 miles to kilometers
This exercise is solved as follows:
 Knowing that 1 mi= 1.609 km
A

$$\frac{1000mi}{1} \times \frac{1.609km}{1mi}$$

conversion factor will be made
 1000 × 1.609= 1609 km
Answer: 1000 mi= 1609 km

29.2 Exercises on Converting Miles to Kilometers

Exercise 1
Converting 30 miles to kilometers

R= 48.27 km

Exercise 2
Converting 200 miles to kilometers

R= 321.8 km

Exercise 3
Converting 500 miles to kilometers

R= 804.5 km

Exercise 4
Converting 5 miles to kilometers

R= 8,045 km

Exercise 5
Converting 10000 miles to kilometers

R= 16090 km

29.3 Examples of converting kilometers to miles

Example 1:
Converting 200 km to miles
This exercise is solved as follows:
Knowing that 1 mi= 1.609 km
A

conversion factor will be made $\dfrac{200km}{1} \times \dfrac{1mi}{1.609km}$

$\dfrac{200}{1.609} = 124.30$ mi

Answer: 200 km= 124.30mi

Example 2:
Converting 500 km to miles
This exercise is solved as follows:
Knowing that 1 mi= 1.609 km
A

conversion factor will be made $\dfrac{500km}{1} \times \dfrac{1mi}{1.609km}$

$\dfrac{500}{1.609} = 310.75$ mi

Answer: 500 km= 310.75 mi

Example 3:
Converting 100 km to miles
This exercise is solved as follows:
Knowing that 1 mi= 1.609 km
A

conversion factor will be made $\dfrac{100km}{1} \times \dfrac{1mi}{1.609km}$

$$\frac{100}{1.609} = 62.15 \text{ mi}$$

Answer:100 km= 62.15 mi

Example 4:
Converting 80 km to miles
This exercise is solved as follows:
 Knowing that 1 mi= 1.609 km
A

$$\frac{80km}{1} \times \frac{1mi}{1.609km}$$

conversion factor will be made

$$\frac{80}{1.609} = 49.72 \text{ mi}$$

Answer: 80 km= 42.72 mi

Example 5:
Converting 25 km to miles
This exercise is solved as follows:
 Knowing that 1 mi= 1.609 km
A

$$\frac{25km}{1} \times \frac{1mi}{1.609km}$$

conversion factor will be made

$$\frac{25}{1.609} = 15.53 \text{ mi}$$

Answer: 25 km= 15.53 mi

29.4 Exercises on converting kilometers to miles

Exercise 1
Converting 800 km to miles

R= 497.20 mi

Exercise 2
Convert 1000 km to miles

R=621.50 mi

Exercise 3
Convert 45 km to miles

R= 27.96 mi

Exercise 4
Convert 20 km to miles

R=12.43mi

Exercise 5
Convert 8 km to miles

R= 4.97 mi

Chapter 30: Conversion from ft to meters and conversion from meters to ft
Examples of conversion from ft to meters

Example 1:
Converting 5 ft to meters
This exercise is solved as follows:
Knowing that 1 m= 3.28 ft
A

conversion factor will be made $\dfrac{5ft}{1} \times \dfrac{1m}{3.28ft}$

$\dfrac{5}{3.28} = 1.52$ m

Answer: 5 ft = 1.52m

Example 2:
Converting 100 ft to meters
This exercise is solved as follows:
Knowing that 1 m= 3.28 ft
A

conversion factor will be made $\dfrac{100ft}{1} \times \dfrac{1m}{3.28ft}$

$$\frac{100}{3.28} = 30.48 \text{ m}$$

Answer: 100 ft= 30.48m

Example 3:
Converting 50 ft to meters
This exercise is solved as follows:
Knowing that 1 m= 3.28 ft
A

conversion factor will be made $\dfrac{50ft}{1} \times \dfrac{1m}{3.28ft}$

$$\frac{50}{3.28} = 15.24 \text{ m}$$

Answer: 50 ft= 15.24m

Example 4:
Converting 10 ft to meters
This exercise is solved as follows:
Knowing that 1 m= 3.28 ft
A

conversion factor will be made $\dfrac{10ft}{1} \times \dfrac{1m}{3.28ft}$

$$\frac{10}{3.28} = 3.04 \text{ m}$$

Answer: 10 ft= 3.04m

Example 5:
Converting 30 ft to meters
This exercise is solved as follows:
Knowing that 1 m= 3.28 ft
A

conversion factor will be made $\dfrac{30ft}{1} \times \dfrac{1m}{3.28ft}$

$$\frac{30}{3.28} = 9.14 \text{ m}$$

Answer: 30 ft= 9.14m

30.2 Exercises for converting ft to meters

Exercise 1
Converting 120 ft to meters

R= 36.58m

Exercise 2
Convert 90 ft to meters

R= 27.43m

Exercise 3
Convert 15 ft to meters

R= 4.57m

Exercise 4
Convert 2 ft to meters

R= 0.90m

Exercise 5
Converting 200 ft to meters

R=60.97m

30.3 Examples of meter to ft conversion

Example 1:
Converting 1000 m to ft
This exercise is solved as follows:
Knowing that 1 m= 3.28 ft
A conversion factor will be made

$$\frac{1000m}{1} \times \frac{3.28ft}{1m}$$

1000 × 3.28= 3280 ft
Answer: 1000 m= 3280 ft

Example 2:
Converting 100 m to ft
This exercise is solved as follows:
Knowing that 1 m= 3.28 ft

A

$$\frac{100m}{1} \times \frac{3.28ft}{1m}$$

conversion factor will be made
100 × 3.28= 328 ft
Answer: 100 m= 328 ft

Example 3:
Converting 500 m to ft
This exercise is solved as follows:
Knowing that 1 m= 3.28 ft
A

$$\frac{500m}{1} \times \frac{3.28ft}{1m}$$

conversion factor will be made
500× 3.28= 1640 ft
Answer: 500 m= 1640 ft

Example 4:
Converting 25 m to ft
This exercise is solved as follows:
Knowing that 1 m= 3.28 ft
A

$$\frac{25m}{1} \times \frac{3.28ft}{1m}$$

conversion factor will be made
25 × 3.28= 82 ft
Answer: 25 m= 82 ft

30.4 Exercises on converting meters to ft

Exercise 1
Convert 10 m to ft

R= 32.8 ft

Exercise 2
Convert 20 m to ft

R= 65.6 ft

Exercise 3
Convert 120 m to ft

R= 393.6ft

Exercise 4
Convert 800 m to ft

R= 2624 ft

Exercise 5
Convert 5000 m to ft

R= 16400 ft

Thank you very much, I hope this book will be useful to you.

If you are a student or person with visual impairment, I would like to give you the following message, continue with all your goals and do not let visual impairment defeat you.

If you are a teacher of mathematics or any other subject, I would like to give you the following message; Help your visually impaired student to learn better and in order to take that step you need to take trainings so that you can successfully apply inclusion, remember that inclusion is not only to include visually impaired students in regular schools as sheep but to give them all the necessary tools.

www.ingramcontent.com/pod-product-compliance
Lightning Source LLC
Chambersburg PA
CBHW070618220526
45466CB00001B/45